P9-EMP-002

DIRT

WILLIAM
BRYANT LOGAN

Riverhead Books
New York
1995

DIRT

The Ecstatic Skin of the Earth

Riverhead Books
a division of G. P. Putnam's Sons
Publishers Since 1838
200 Madison Avenue
New York, N.Y. 10016

Published simultaneously in Canada.

Library of Congress Cataloging-in-Publication Data

Logan, William Bryant.
Dirt : the ecstatic skin of the earth / William Bryant Logan.
p. cm.
ISBN 1-57322-004-3
1. Soils. 2. Soil—Social aspects. I. Title.
S591.L568 1995 95-7317 CIP
631.4—dc20

BOOK DESIGN BY JUDITH STAGNITTO-ABBATE

Printed in the United States of America
1 3 5 7 9 10 8 6 4 2

This book is printed on recycled paper.

For Hans

ACKNOWLEDGMENTS

THIS BOOK IS OWED TO SO MANY PEOPLE, it is hard to know where to start, except to say that none of its flaws or errors are the fault of anyone but myself.

I owe an immense amount to Hans Jenny, who advised and inspired me. No one has ever known the soil better than he did. The Very Rev. James Parks Morton, dean of the Cathedral of St. John the Divine, made this book possible almost a decade ago, when he invited me to become a writer in residence. He has been an example to me of work pursued tirelessly and with deep good cheer. He is the "friend" referred to in the first essay, entitled "Stardust." The compost men Will Brinton and Clark Gregory were both extremely patient with me, explaining their work and correcting my mistakes. Many, many scientists shared the fruits of their research with me, and I am grateful to them for their illuminating work.

My agent, Molly Friedrich, showed a persistence and loyalty that was little short of miraculous, through good times and bad. My editor, Amy Hertz, has been a great supporter and a pleasure to work with.

Also, I want to thank everyone—friend, acquaintance, and stranger—who sent me little bags of soil over the long course of this project. Getting a new batch of dirt always lifted my flagging spirits and reminded me of Hans Jenny's dictum that every soil is an individual. My son Sam was a wonderful experimenter and helped to open my eyes with fewer preconceptions. My daughter Eliza, who had not even been born when this book was begun, now loves to dig beside me and to sing while she digs.

There were others whose advice and example were crucial to me, and though I cannot name them all, I thank them all.

CONTENTS

Only the mistakes are ours. In this world, beauty is common.

—Jorge Luis Borges

In that hour, weary of life, men will no longer regard the world as a worthy object of their admiration and reverence. This All, which is a good thing, the best that can be seen in the past, the present and the future, will be in danger of perishing; men will esteem it a burden; and thenceforward they will despise and no longer cherish this whole of the universe, incomparable work of God, glorious construction, good creation made up of an infinite diversity of forms, instrument of the will of God who, without envy, pours forth his favour on all his work, in which is assembled in one whole, in a harmonious diversity, all that can be seen that is worthy of reverence, praise and love. For darkness will be preferred to light; it will be thought better to die than to live; none will raise his eyes towards heaven; the pious man will be thought mad, the impious, wise; the frenzied will be thought brave, the worst criminal a good man. The soul and all the beliefs attached to it, according to which the soul is immortal by nature or foresees that it can obtain immortality as I have taught you—this will be laughed at and thought nonsense.

—FROM ASCLEPIUS; TRANSLATED FROM THE ITALIAN TRANSLATION BY MARSILIO FICINO

DIRT

CLYDE'S PICKUP

A YEAR AGO, CLYDE FELL OFF A SCAFFOLD. He is a big, black Texan who wears a worn straw cowboy hat and the same jeans every day and boots that he seems to have been born in.

He had assembled the scaffold out in front of the Great Portal of the Cathedral of St. John the Divine in New York City. He was to check the mortar on some limestone blocks high overhead, lest they come down some Easter morn and clean the bishop's clock while he stood waiting to enter church in the role of the risen Christ. Somehow, Clyde fell more than forty feet onto gray limestone steps.

While he convalesced in the hospital, his black Chevy pickup truck sat unused and unmoved in a space at the head of the driveway, under a maple tree. The cathedral's urban pigeon corps has made great cloudlike white patches of droppings all over it. Inside the cab, between the dashboard and the windshield, are stuffed sheafs and wads of notes, instructions, registrations, a box of toothpicks, cassette tapes, catalogues, the tops of commuter coffee cups, stirrers, newspaper clippings, chopsticks, and saw blades. This is Clyde's filing system.

In the back of the pickup lies a rough pile of pigeon-spattered sawed logs, along with a generation of fallen leaves, a broken fanbelt, an empty yellow antifreeze container, numbers of Styrofoam cups, a boasted stone, a rusting can of Super Stripe traffic paint, a few discarded service leaflets and menus for Chinese food, a ticket that reads "admit one," and a vintage book of diocesan records with advertisements for long-vanished vestment stores and Episcopal schools, their lettering now half eaten by mold.

Out of these leavings a forest is growing. Not on the ground. Not beside the truck. Right in the back of it. The lobes of maple leaves are sharpening as their seedlings sprout, a light and glossy green. The red-stemmed and three-leaved poison ivy is showing its amazing skill at growing out of any slightly shady bit of dead wood. Seeds of albizia have somehow blown from the one little understory specimen halfway down the street and taken hold in the back of the truck. All of this is happening in New York City, fifty yards from Amsterdam Avenue, where the eighteen-wheel trucks whiz by.

Inside the cab, its windows closed, the dust gathers, the rip in the upholstery creeps infinitesimally toward the back of the seat, the papers yellow and curl. Nothing grows. The FAN and the HOT-COLD levers remain where they have been for a year. PARK-R-D-2-1. Nothing is happening, because the motor is not running.

But in the back of the truck, open to the air and the wet, Great Nature's motor is emphatically running. Left out in the rain, the diocesan book has a sprout in it. The tilted coffee cup has filled with leaf compost, dots of pigeon shit, and wood mold, and albizia are growing in it. The old black Chevy is alive.

Wherever there are decay and repose, there begins to be soil. It would be hard to imagine a more improbable set of ingredients, but even a truck can become dirt.

How can I stand on the ground every day and not feel its power? How can I live my life stepping on this stuff and not wonder at it? Science says that an acre of soil produces one horsepower every day. But you could pour gasoline all over the ground forever and never see it sprout maple trees. Even a truck turns to soil. Even an old black pickup.

Recently I have been reading Exodus, wondering about Moses and the burning bush. Moses, it is written, "turns aside to see a wonder," a bush that burns but is not consumed. Throughout my life, I had thought this a ridiculous passage. Why should God get Moses' attention by such outlandish means? I mean, why couldn't He just have boomed, "Hey, Moses!" the way He would later call to the great king, "Hey, Samuel!"

Now I know why. The truth, when really perceived and not simply described, is always a wonder. Moses does not see a technicolor fantasy. He sees the bush as it really is. He sees the bush as all bushes *actually* are.

There is in biology a formula called "the equation of burning." It is one of the fundamental pair of equations by which all organic life subsists. The other one, "the equation of photosynthesis," describes the way that plants make foods out of sunlight, carbon dioxide, and water. The equation of burning describes how plants (and animals) unlock the stored sunlight and turn it into the heat energy that fuels their motion, their feeling, their thought, or whatever their living consists of.

All that is living burns. This is the fundamental fact of nature. And Moses saw it with his two eyes, directly. That glimpse of the real world—of the world as it is known to God—is not a world of isolate things, but of processes in concert.

God tells Moses, "Take off your shoes, because the ground where you are standing is holy ground." He is asking Moses to experience in his own body what the burning bush experiences: a living connection between heaven and earth, the life that stretches out like taffy beween our father the sun and our mother the earth. If you do not believe this, take off your shoes and stand in the grass or in the sand or in the dirt.

I

The Ecstatic Skin of the Earth

STARDUST

YOU ARE ABOUT TO READ A LOT ABOUT DIRT, which no one knows very much about. We don't even know the real etymology of the word. It's that stuff that won't come off a collar. It's what smells in a compost heap. It's what blows around on the floor or makes the sheets feel gritty and slick. It's dog turds. It's the stuff we walk on, and traditionally, it is where people plant crops, though nowadays of course hydroponics is supposed to make that grimy connection a thing of the past.

The truth is that we don't know the first thing about dirt. We don't even know where it comes from. All we can say is that it doesn't come from here. Our own sun is too young and cool to manufacture any element heavier than helium. Helium is number two on the periodic table, leaving some ninety elements on earth that were not even made in our solar system. Uranium and plutonium, the heaviest elements that occur in nature, can be forged only in an exploding star, a supernova.

"We are all stardust," says a friend of mine. He understates the case. In fact, *everything* is stardust.

How does the stardust get here? "The interstellar medium is full of dust," says University of Washington astonomer Donald Brownlee. "It's what makes the dark line across the center of the Milky Way." It is a line of dirt perhaps 65,200 light-years across, and 3.832×10^{17} miles long!

There are countless megatons of unknown dirt out there. The ejecta, dejecta, rejecta, and detritus of ruined stars, it floats around the universe until it enters a field of force. Force fields of any kind resemble spiderwebs. Some are concentric, some are spiral, others are like labyrinths. They are traces of the way that a force—magnetic, kinetic, gravitational—acts. Of whatever kind, they tend to catch passersby.

Imagine that the dust from a thousand different exploding stars has gathered along the weak lines of magnetic force in a spidery red nebula. A few light-years away another star explodes. The force of the bang sends out a shock wave that perturbs and twists the magnetic lines of the nebula, creating eddies and whirlpools, exactly like those in a river. In those eddies, the dust begins to gather. As it forms lumps, gravity for the first time becomes stronger than electrical forces. The process feeds upon itself, until spherical masses are formed.

If a very large mass accumulates, it may catch atomic fire and become another sun. If smaller masses form, they may become planets of solid, liquid, and gas. If still smaller, they may become the cold hunks of moons, asteroids, or comets.

Perfectly logical, right? Indeed, just as logical as those experiments in which flat squares of metal are coated with a layer of loose sand. When music is played at them, the grains of sand arrange themselves in simple, concentric shapes, as the otherwise invisible signature of those tones. Perfectly logical, but why does it happen?

Apparently, the universe sorts more or less dense matters for different purposes on different scales. At the scale of the sun, atoms can't exist. Their electrons are blown off and ions constantly radiate from the sun, spreading through the solar system.

At the scale of the moon, matters are the contrary. They are loose and virtually inert. Meteor strikes have been stirring it up for about four billion years, so it is all thoroughly mixed. The solar wind constantly pummels it, embedding helium atoms in the lunar soil.

Who would expect something like the Earth to come between sun and moon? Nobody. The sun is all activity, the moon all passivity, but the Earth is both active and passive. In the course of its evolution, this compacted mass of interplanetary dirt called the Earth, has two primary products: soil and atmosphere.

From this point of view, life on Earth is a kind of machine for making soils and atmospheres. Volcanoes disgorge oxygen-poor, virgin mineral materials from deep inside the crust. Rising into the stratosphere, they pick up oxygen atoms and fall with them to Earth. Meanwhile, the green plants and the blue-green algae—where did *they* come from?—convert sunlight and carbon dioxide into food and oxygen. Animals—*what are they?*—take the oxygen and convert it back into carbon dioxide. The result is a vast cell—the Earth itself, hurtling through space with its envelope of atmosphere and dust—that regulates the inflow of solar hydrogen and electronic energy and the outflow of energy made by combining these with the heavier elements of Earth.

As the beings that make up organic life continue to exist, evolve, and cover the Earth, they create a rich, stable atmosphere and rich deep soils. Only here on Earth does stardust engage in this extraordinary array of self-organizing behaviors. Only here on Earth does it perform the ceremony of continually creating an atmosphere. Even our neighbor Mars cannot do so, because its crust has become an unbroken solid and volcanoes cannot exist there. Mars has no means of self-renewal, no connection between its depth and surface.

Why is Earth's dirt special? To a scientific mind, it is hard to admit that we don't, and possibly can't, know the answer. Answerless questions are the best kind. What's more, it seems that things that can't be figured out can still be *seen* to be true. Confession, not of "sin" but of ignorance, and meditation, not on some mantra but on the created, yield results that are different from analysis, and much more powerful.

I confess that I do not know, and I begin to meditate upon the wonderful construction of this world. A different part of the mind seems to open. Things that once seemed trivial now assume their right importance, and coincidence reveals its purpose. "Faith," as the early Christian says, does not mean "belief." It means the substance of things hoped for and the evidence of things unseen.

It is all impossible, after all: the Earth, the dirt, and all these things that we do not know and that we did not make.

"Let us worship the Lord in the beauty of holiness," says the psalm. What is beauty? Beauty is a sum not reducible to its parts. It is a perception of harmony in variety. What is worship? To worship means not to figure out, not to analyze, not to pin down like a dried butterfly on a grid, but to value. Deeply to value.

THE FIRST SOIL

La fixité du milieu interieur est la condition de la vie libre.
[To contain the inside is to make free life possible.]

—CLAUDE BERNARD

THERE MIGHT HAVE BEEN A MOMENT when the Earth was perfect and unblemished. This moment would have occurred maybe four and a half billion years ago, when a thin crust congealed on the surface, creating a vast membrane that separated inside (the hot core) from outside (outer space). The differential thus inaugurated immediately caused eruptions, the hot interior punching volcanic holes through the membrane to vent its agitated nitrogen, carbon dioxide, and water vapor on the surface. The roiling interior fluids pushed up humps and ridges, cones and mounds, while meteor impacts from the outside pressed out concentric mountain ranges with diameters in the hundreds of miles. The Earth must have looked like a dead body in the first throes of decomposition.

For perhaps half a billion years, the place was too hot for life. Water remained as vapor in an atmosphere rich in carbon dioxide, formaldehyde, neon, and cyanide. Then, as the Earth began to cool, it rained for perhaps twelve thousand years without stopping, helping to create the first seas.

The surface of the Earth was black and white then, rather like the aseptic photos of the surface of the moon. The browns, the yellows, the reds, and the oranges that we associate with soil and rocks did not exist, because all of them are the signs of oxidized iron. In that early time, the iron was all dissolved in the water—it dissolved as easily as salt does today.

The falling rain scoured the fresh curves of the land, finding the weak spots and the fissures. Sheets of flow found the lines of weakness, concentrating into torrents and cataracts that swept fragments of the mineral elements into the new seas.

The sea was the proto-soil, where Earth, air, water, and the solar fire met for the first time. It was an inverse soil, you might say, with the liquid element providing the matrix for the mineral salts and for dissolved gases, a role that the mineral elements would later come to play. But from a certain point of view, all Earth's later history is a consequence of that first mixing. In that sense, life is the story of bodies that learned to contain the sea.

One morning in the Archean Era, an assembly of chemical compounds, possibly a clay, obeying some divine suggestion, threw an envelope around itself and began to live. It now had an inside and an outside. The envelope regulated the flow of salts from one to the other, and back again, making the selective work of digestion possible.

In the three billion years since, organisms have merely perfected this design. Though those organisms were microscopic creatures that probably ate iron sulfides, while we are masterpieces of bilateral symmetry, capable of eating everything from steaks to nasturtium flowers, we and the plants and the microbes are all containers for a fluid that is very like the sea. Ours is red, because we use an iron compound called "haem" to multiply our ability to extract oxygen from the air. A plant's is green, because it uses a magnesium compound instead of iron, to extract carbon dioxide. A sea squirt's is colorless, and will run out of its body if you expose it for too long to the air. But all of them share a need for a very narrow spectrum of salt concentrations, derived from the first weathering of rocks in the Archean Era.

The bodies of plants and animals concentrate the salts of the Earth, selecting those ions which are of use to metabolism. Silicon and aluminum, though they make up a large fraction of the soil, are present only in trace quantities in humans and plants. Calcium, magnesium, potassium, phosphorus, and sulphur, by contrast, exist in increasing concentrations from soil to plant to man. The ratio for calcium is 1:8:40, for phosphorus 1:140:200.

Eon by eon, the creatures of the Earth have moved toward what biochemists call "osmotic independence," that is, the ability to carry these salty fluids freely through the open air. They can walk around without drying out. The right sort of impermeable protective covering is needed—chitinous like an insect's plates, or, better, keratinous, like fur—and the right sort of pumps, particularly excretory pumps and circulatory pumps, as means of getting salts in and out and moving them about.

The floating and the bottom-dwelling invertebrates of the seas are memorials to the earliest strategies for achieving this freedom, but though they acquired a skin, they did not acquire the ability to move under their own power. In a sense, they were and have remained cells in the vast organism of the ocean, which moves them at will. Ocean currents are furrows along which its life glides.

Freedom of movement comes with freedom from the rhythms of the sea. Even in our bodies, the lymph and the lung fluids are controlled by tidal forces, but our pumps by their constant work give us a range of freedom of movement that was previously impossible. The heart's tissue is created to beat, and continues to do so even after it is separated from the body.

Life did not crawl out of the sea onto the land; it oozed from the sea into the land, the organic acids of its excretions joining with the carbonic acid of the rainfall to create the first soft mantle of soil on the Earth. Maybe two billion years ago, the cyanobacteria began to use the sunlight to make sugars, excreting oxygen. They were green or brown, and their scum spread into lagoons, up rivers. The oxygen reacted with iron, and for the first time there were orange, yellow, and brown colors in the earth.

This was the world's first bloom. Scientists often attach to these colonies of free-living organisms the unattractive moniker "algal mats." Dense symbiotic colonies of cyanobacteria, fungi, and molds formed crusts that held dissolved minerals in place, preventing them from washing into the sea. They probably had all the various beauty of the orange, the red, the yellow-green, the filamentous, and the ramified lichens that cling to rock surfaces today. More precisely, they may have resembled the similar colonies called "cryptogamic crusts" that survive in arid climates today, keeping enough moisture among them to support life in a hostile climate.

On the first soils in the first season of the Earth, individuation began. Separated now by a solid matrix instead of by the shifting waters of the sea, organisms found it possible to wave without moving away. Things could stay in one place. Burning high-powered oxygen, they would soon be able to wave one way while moving in a completely different direction.

Strangely enough, when you look for a creature to match the range of motion of the human hand, you find yourself back with the wiggling orange filaments of fungi and the gesture of acclamation of a spreading bacterial branch.

HUMUS

THE TEXTBOOKS WILL TELL YOU that humus is "deeply altered, black organic matter" lying in the top layer of the soil. But I only grasped what humus is about when my friend Pamela Morton showed me a picture of it.

Pamela spends parts of her summers in boreal Canada and has long been fascinated with the forest floor, where bits of blackening twigs, needles, bark, and the carcasses of small creatures decay. Humus is this organic part of the soil, the final residue of those matted leaves and cold bodies, intercalated, lapped, melding, losing their distinction, dry on top but inside turning the shades of brown and black that we associate with rain-wet wood or a man's study. The deeper you go, the blacker it gets, and the fewer of the bits survive intact. A few inches down, it is pure black acrid matter having a texture like a cross between cotton candy and damp sawdust. This is the stuff from which all life on the land is born.

Pamela wanted to make a picture that would show its power. She began to experiment, making homemade paper out of leaf litter and deriving vegetable dyes from the material of the forest floor. She got not only deep blacks, but a range of browns, oranges, yellows, and reds, with a range of blue from robin's egg to midnight.

All this was hidden in the substances themselves. She made of it a small collage, no bigger than the cover of this book, showing matrices of shifting color patterns, with a deep black enclosing them all. I have never seen an image that more dramatically showed the huge energy contained in humus. It was a dynamo of pure hues shining out of the blackness.

It set me to wondering about humus.

Humus, human. The dictionaries say there is a connection between the words, but they don't elaborate. What does the root *hum-* mean?

It must have to do with *humble,* or with *humilis, humiliate.* Those words come from roots meaning "of the ground, lowly." But humus does not refer to the ground itself. It refers to the end product of decaying litter and dead creatures. It also has to do with being *humorous,* that is, in the original meaning, "wet." Both people and humus are wet inside.

Wetness is opportunity. It represents the openness of nature to what falls from heaven. As Meister Eckhart put it, the humble man is "he who is watered with grace."

The processes of growth, decay, feeding, digestion, excretion, attack, and repulsion all need wetness and generate heat. To understand them, you have to study the interconnections, not the essences. You have to put physiology first, not molecular chemistry.

For more than a century, chemists have been trying to answer the question, What is humus? And to this date, no one knows. Probably, no one will ever know. Even Hans Jenny, who knew more about the soil than anyone, remarked with a sigh, "Humus is imperfectly understood." Every time you attempt to break it down into its basic components you get acids of a slightly different nature. All of them have similar properties, for example, a carbon-to-nitrogen ratio of 10:1, but none of them are chemically alike. In fact, as soil scientist Dr. James Rice puts it, "It is very possible that no two humus molecules are or have ever been alike." Like snowflakes or people.

This makes it virtually impossible to apply to humus the quantitative experimental methods upon which modern agricultural chemistry is based. In fact, to establish that chemistry, the great nineteenth-century chemist Justus von Liebig had to try to descredit the then-prevailing theory that humus was the chief source of plant nutrients.

Liebig established that to grow a crop successfully, you need only supply the requisite mineral elements plus nitrogen in a sufficient quantity for the plant in question. Whether these substances are drawn from humus or not is immaterial, he argued. And he ridiculed the humus theory for its insistence that all plant carbon is derived from humus. In this respect, he was right, since most plant carbon is ultimately derived from the carbon dioxide of the air.

Today we still do not know what humus is, but we know a little about what it does. Some of it releases nitrogen and trace elements for the reuse of plants; all of it nourishes the microbes, which decompose it and whose bodies add to its substance. The microbes, furthermore, secrete sticky substances that help bind humus and clay together into stable aggregates. We know, too, that, unique among biological substances, humus resists the processes of microbial decay, so that it can remain in the soil, sometimes for ten thousand years or more. And we know that this is a good thing, because it can hold mineral nutrients for a plant's use twice as well as the best clays. We are also aware that stable humus helps a porous soil to hold more water, and a heavy soil to hold less water. (A pound of sand absorbs one fourth pound of water; a pound of humus absorbs two pounds.) And finally, we see that in this ensemble of properties, it is the habitat for a diverse microflora and microfauna that tend to suppress or eliminate disease organisms, in order to continue their own robust lives.

Radical disorder is the key to the functions of humus. At the molecular level, it may indeed be the most disordered material on Earth. No two molecules of humus may be alike. Though no one has difficulty recognizing a humus molecule, it is quite likely unique, because it works upon fractal principles. Simple geometries define any given part of it, but the modes for the combining of these shapes produce a vast array of different manifestations at different scales. For humus, similarity is rampant, but identity nonexistent.

Neither humus nor humans are humble at all. We are audacious, like nature herself. We are wet, fecund, protean, dangerous. When we start to comprehend this in widening circles of the world, we know something worth knowing. We know that we must become responsible.

SAINT PHOCAS AS FERTILIZER

FAR TO THE EAST OF CONSTANTINOPLE, near the town of Sinope on a thumb-shaped peninsula sticking out into the Black Sea, lived Phocas the Gardener—*animum simplicem, hospitalem,* a simple soul and hospitable, as a fourth-century hagiographer said of him. Nobody knows when he lived, but many know how he died. He became the Christian patron saint of the garden, because he composted himself.

This is the story.

A persecution arose against the Christians. Phocas was denounced as one of the banned sect, and a pair of Roman soldiers were dispatched to find him. "Don't worry about evidence, don't worry about a trial, don't worry about a confession," they were told. "When you find him, kill him."

The soldiers had hard traveling to reach Sinope, over thickly wooded mountains that to this day isolate the coastal port from the Turkish interior. Already, in the early fourth century, the town had declined from its peak of prosperity a century before, when it had been the terminus of a caravan route to and from the interior steppes and the preferred provider of the ochre derived from so-called "synopic" red earths. But the caravans were no more. Then as now, dangerous roads traversing passes at over four thousand feet in elevation had made the trip a chancy one, scarcely worth the rewards it could bring.

So the soldiers were not in a good mood as they descended the steeps, catching sight of the broad uplifted thumb of Sinope's little peninsula. It was a surprisingly fertile land. Its igneous bedrock, covered with fine coastal alluvium, had been uplifted by tectonic forces, forming both a sheltered port and a fine coastside farming belt, with a mild climate and temperate rains scattered throughout the year. Phocas had not chosen his homestead badly. In the region, he could easily have grown everything from grains to grapes, from filbert nuts to cabbages.

The weary pair trudged northeast onto the isthmus, bearing for the town, where they might ask directions to Phocas's house. Night overtook them near a farmhouse, where a pleasant man of no particular age greeted them.

"You are tired," he said. "You must stay the night with me." When they began to protest that they had a pressing errand, he responded definitely, "I won't take no for an answer."

Over the evening meal, the soldiers told their host that they were looking for a dangerous man named Phocas. The host assured them that he knew the man very well and could quickly point him out to them. As they were so weary from their journey, however, he suggested that first they rest for the night.

While his guests slept, the host took his spade and dug a broad, deep hole in the middle of his garden. Come morning, the man fed the soldiers a hearty breakfast. Then, he told them that they might capture Phocas whenever they wished. He knew just where to find him.

To have come out of the wilderness to such a welcome had been pleasure enough. Now, this kind host had simplified their duty for them, saving them hours, maybe even days of asking hard questions, getting evasive answers among a surly, uncooperative people, probably half of them Christians and the other half brigands.

They thanked him for his thoughtfulness and asked to be taken to Phocas, wherever he might be.

"He is here," said the farmer. "I myself am the man."

Their jaws dropped. Their minds froze. What a trap! How could they behead this man, who'd treated them so well?

But Phocas led them to the hole he'd dug in the garden, and there, with his consent, they chopped his head off.

That's how the story goes, but to take it a step further, we must imagine the soldiers carefully covering the hole with soil, as well they might have. It was the least they could do for a man who'd taken such good care of them. And we must imagine the thoroughness of Phocas's simple and hospitable soul, which took such care to return to the garden the body that had taken sustenance from it.

The fungi colonized it first, hydrolizing the tissues without disturbing the form. Then the white worms and the maggots and the mites took over, breaking off larger chunks, ingesting these, themselves defecating and dying. And this increasingly diverse pile of remains was attacked by wave after wave of further bacteria and fungi, until at last Phocas's mortal part had been completely oxidized.

The nitrogen of his protein and nucleic acids was fixed as nitrate or released as ammonia or other nitrous gases. His carbon went largely into the air as CO_2. Iron and phosphorus and other elements remained in the soil combined with oxygen. In short, his tissues rejoined the cycles of elements from which they had been briefly extracted.

The strongest and most lasting thing about Phocas was his soul. The kindness that he did has kept him alive for almost two thousand years now, while the carbon and nitrogen that once held his body together have been recycled billions of times. (If you want to do the calculation, figure that oxygen takes a seventh of a second to recombine, and carbon dioxide takes one five-thousandth.)

The hagiographer Butler claims that Phocas "found in his garden an instructive book and an inexhaustible fund of meditation." I would say rather that he found lessons in simplicity, economy, and hospitality, the three virtues most often found in gardeners. There is no need to suggest anything about Eden. If Phocas had meditated on any story from the Bible, it might have been on the parable of the seed that was planted on stony ground, the other planted where the birds could take it, and the third planted in a good soil so that it brought forth fruit. To make that soil, Phocas did not spare even his own body.

Hospitality is the fundamental virtue of the soil. It makes room. It shares. It neutralizes poisons. And so it heals. This is what the soil teaches: If you want to be remembered, give yourself away.

SWEET AND SOUR SOILS

IT USED TO BE THAT A GOOD FARMER COULD TELL a lot about his soil by rolling a lump of it around in his mouth. I once spent a week trying to find somebody who still did that. Not a prayer. Not in Texas, nor Vermont, nor Kentucky, nor California, nor western New York. Everybody knew somebody who once did it, but nobody could quite remember the name of the fellow.

Finally, I found Bill Wolf, president of a company called Necessary Organics. Wolf had chewed soil himself up to a few years previously, when his doctor had strictly forbidden him. Soil contains bad bugs as well as good ones, and the physician did not want to have to sort them out in Wolf's guts.

But in the days when he chawed, Bill could tell acid from alkaline by the fizz of the soil in his mouth. A very acid soil would crackle like those sour candies that kids eat, and it had the sharp taste of a citrus drink. A neutral soil didn't fizz and it had the odor and flavor of the soil's humus, caused by little creatures called "actinomycetes." An alkaline soil tasted chalky and coated the tongue.

For most plants, neither of the extremes is palatable, any more than the fizz or the chalk was to Wolf. Here is the graphic and immediate reason that pH actually is important.

If you mention the word "soil" to a gardener, the first question is liable to be, "What is its pH?" In answer, you give a number between 0 and 14. Whatever the figure, it has the force of a talisman.

The pH value is one of the few instances in daily life where moderation is graphically praised. In almost any sporting event, one seeks the highest (football, baseball) or the lowest (golf, track) number. But if you tell a gardener that his soil's pH is 14 or 0, he will keel over in a faint. Even 3 or 8 is very unpleasant to hear. No, a gardener cannot be leased unless the number is greater than 5 and less than 7.5. The middle terms are where fertility lie.

In the microcosm, this principle is represented in the play of acids and bases, by means of which rocks are broken, foods mobilized, digestion accomplished, excretion stimulated, growth catalyzed, destruction indicated, and new life spawned. pH is simply a way to talk about this game.

pH measures the activity or concentration of hydrogen ions in a substance. Scientists being devious creatures, however, they arranged the scale that measures pH so that a small number means lots of hydrogen, while a large one means little hydrogen. The scale runs from 0 to 14, and each whole number change is equivalent to a tenfold decrease or increase.

An acid has a greater concentration of hydrogen ions (which are the same as a single proton), a base has less. Acids and bases react together to make salts. It might be said that hydrogen has two personalities. In the sun it is the source of the incandescence that makes all life possible. In the sublunary world, it is the chief substance that regulates the exchange of mineral substances. Ecologists say that Earth has a surprisingly high pH. What does this mean?

A cell full of saltwater or my coursing salt blood or the plasma in a plant are all media for the exchange of the bases calcium, magnesium, and potassium. It is these elements whose circulation balances the chemistry of the Earth, raising the pH. Their dance with the acid ions of hydrogen regulates most of the processes by which cells make exchanges.

Consider their interaction in the soil. When the rains come, the rainwater brings hydrogen to the particles of clays and humus. The highly charged hydrogen ions muscle onto the surfaces, pushing the bases calcium, potassium, and magnesium into the soil solution. From there, either the bases will leach down and out of the soil or plants will take them up.

The bases bloom into solution, while the soil particles themselves turn acid. The roots of the plants sweat a mucus that captures clays and ions. Releasing a stream of hydrogen ions themselves, the roots make space on their electrical edges to receive still more bases. The opposite of what is happening to the soil particles happens to the roots, so that they dance like a man and woman in the play of action and reaction, forward step to backward step, backward step to forward step. An alternating current of ions flows into roots' cells along these infinitesimal, reciprocating channels; a channel of hydrogen flows out, making the surrounding soil medium more acid still.

The soil and a tree transform each other, through the medium of the exchange that happens in the roots. A forty-five-year-old shore pine growing in isolation on a hill of shelly sand causes the pH of the soil under its canopy to fall by 1.5 points (that is, the hydrogen ion concentration was raised 150-fold) in comparison with the fresh sand around it. At the same time, however, the soil doubled its ability to exchange bases, and developed a black litterfall horizon rich in nitrogen.

All this depends upon the life-giving osmotic gradient, the difference in salt concentration that allows bases to flow from a place of lesser concentration (the soil) to greater concentration (the inside of the cell). When the flow is reversed by an excess of salts, mineral-laden water flows out of the root cells, and they die of dehydration. This is precisely what happens when a dog pees on the lawn.

You have only to remember how sick and cramped you felt when after vigorous exercise in the hot sun, you failed to replace the salts you had sweated out. Drinking water only increased your nausea, because the salt was still on the wrong side of the cell walls.

The agricultural cure for both acid and salted soils is to apply natural materials that are rich in bases. Limes (calcium carbonate) and gypsum (calcium sulfate) react with the hydrogen of acid clays and the sodium of salt-clays, respectively, sending the hydrogen and sodium into the soil solution, improving the soil's structure, and capturing the nutritious calcium. Some of our lesser-known heroes, the eighteenth-century statesman Edmund Ruffin and the twentieth-century soil scientist E. W. Hilgard, were responsible for the development of these all-important soil cures.

In the autumn, the tide of bases climbs into the finest pores of plants—the tissues of flowers, leaves, and fruits—whence it falls to the ground. Within a day, almost half the fallen tissue has been digested by the microbes and invertebrates dwelling on the ground. The acidity of the soil recedes, and it prepares for its slow, neutral winter life, making an equilibrated medium to protect the roots until the spring.

THE SAND DROWNS THE SEA, THE SEA TAKES THE SAND

THE WEALTH OF AMERICA IS BASED on the black soils of its prairies, whose broad flat expanses are often compared to a sea. When I was in Chase County, Kansas, I perceived directly that this was no empty metaphor.

I'd gone to pee against a fencerow by the highway, making use of my mammalian ability to jettison concentrated, excess salts. As I stood looking into the prairie, I noticed in a nearby gully that there was a thick white layer beneath about a foot and a half of black loam.

Ducking under the barbed wire, I walked along the gully, until I began to hear a crunching beneath my feet. Littering the ground were whitened hunks of coral, the shells of a kind of horned clam, and button-sized wheels of calcium, the remains of sea creatures that had died on this spot more than 250 million years ago. Quite literally, this soil—now half a continent away from the roiling surf of either the Atlantic or Pacific oceans—was once a seabed.

The sea can be a source of both fertility and sterility. A man in Malta once thought he would create a saltworks by digging shallow basins atop a headland fifty feet high and supplying them by drawing up water through a hole he drilled down to the level of the sea caves. The water in these pools simply drained away through the pores in the rock, so he never got his salt. Worse, when the driven waves pressed into the sea caves beneath, they caused a jet of water to shoot out of the hole sixty feet into the air, spraying crops of the whole neighborhood and damaging or destroying them. They tried to stop up the hole with stones, but every year, the winter surf blew the rocks out again.

Sodium salt is one of the worst poisons that can affect a soil, yet the best agricultural soils in the world exist at the boundary between the sea and the land. The Dutch polders, having been formed by rivers delivering their alluvial sediment to the sea, are matchlessly fertile, once they have been leached of this sodium. Thirteen of the sixteen elements that plants need to have are derived from the minerals in soil, particularly from rich alluvium. In Lincolnshire, British farmers fertilized their fields by "warping," that is, by allowing the seawater to cover them at high tide, giving them a coat of rich slime derived from the sediments of river deltas.

So valuable are the soils deposited at the seaside that governments have elaborated incredible hypotheses for staking a claim to them. Napoleon, for example, reasoned that since the entire Netherlands were really a deposit of the Rhine River, therefore, they belonged to whoever owned the sources of the Rhine. In other words, they belonged to Napoleon.

The shoreline, this place of pummeling, where even the air may be poison to a crop plant, is the probable first home of all living things. In the beaches and estuaries, the tidal flats and saltmarshes, the submarine canyons that mark former river bottoms, the mangrove swamps and the flocculated benthic soils that form where the deltas drop their load, is the most diverse habitat on Earth.

It is as though the Earth had tried to produce as much variety as possible in the hope that some of it would make the transition from water to air. A single tidepool on the Pacific coast may contain several thousand different species of invertebrate organisms alone, from the sand-colored brittle star beneath a rock, to the bubble-gum-pink nudibranch called Hopkin's Rose, clinging to a frond of kelp.

In the littoral zone the Earth performed its great experiments with maintaining salts in the body. Here were the first creatures that did not need to be surrounded by seawater. Essentially, because the tide rose and fell, they had to be prepared for both wet and dry. Crabs and other crustaceans, especially the isopods, were the early possessors of a tough cuticle that allowed them to store water in the dry periods, and so to store up the salts that they would need for their metabolism.

The shore is a laboratory of salts. The kelps of the littoral zone have been known to be good fertilizer since the fourth century at least. In Elizabethan times, they were called "the poor man's manure."

But kelp is not the only creature that takes up the salts brought to the sea. The sodium, the calcium, the magnesium, the potassium, and the bicarbonates combine chemically into either limestone or dolomite, forming seabottom layers that someday, like the sea-formed limestones of the Midwestern soils, will make fertile soils.

VIRGIN SOIL

SOME AGRONOMISTS, trying to make a good thing out of global warming, suggest that with rising temperatures we will soon be able to farm northward into Canada and Siberia, and higher on the slopes of mountains. The changes in climate, they predict, will stimulate immediate fresh soil building.

It isn't so. The virgin, never-cultivated soils of the north and the heights are stiff and stony. Even though reaction rates double with each ten-degree rise in temperature, a fertile soil does not appear overnight. The geologist William Fyfe has studied this question, observing volcanoes at different latitudes to see how fast their fresh lava turns to soil. Even in Hawaii, he finds, it takes at least one thousand years for the first centimeter of fresh soil to form. So if the Northwest Territories turned into Mauna Kea, we would still have to wait perhaps ten times ten centuries to plant the wheat.

Virgins are particular. They call forth the husbands appropriate to them. In the tundra of the Northwest Territory, it takes 120 square miles to feed a single human being. Not grain or corn, but caribou, is the transformer that converts the energy of this thin soil into food for a man.

A tundra soil does not look productive. Rainforest soils, on the other hand, dress richly, but only to conceal their underlying poverty. In fact, the better virgin soils there—for example, in parts of the Amazonian rainforest—are in disguise, cloaked not with behemoths, but with slender acai, becacu, and mora trees. The great 150-foot trunks of the acapu, the caju-acu, and the jurana trees stand on far poorer soils. The native peoples know this, choosing sites for their shifting slash-and-burn fields by looking for the slender trees. The *colonos,* recent immigrants brought from the coast under government programs, burn off the *large* trees, assuming falsely that they must have grown on the better soils. To make matters worse, while the natives grow locally adapted cassava in their plots, the *colonos* import nutrient-hungry and ill-adapted rice, corn, and beans. Of the thousands of plots laid out to either side of the Trans-Amazon Highway to a distance of sixty miles, only a small fraction have been claimed.

Even the virgin soils of the temperate forest are delicate creatures. A temperate forest lives largely by recycling its own masses. If you remove the litter, you break the chain, and the nutrient-poor soil beneath cannot supply the deficit.

Consider the growth of a redwood tree on the Northern California coast. On the day Christ was born, the seed falls in sand deposited at the bend of a river. The diving roots send out sugars that attract microbes that feed there, living, dying, and decaying. The young tree begins to drop its foliage, bark, and cones, along with the bodies of insects and other animals that live in it. Organic acids in the soil water leach the iron and aluminum from the sand and begin to carry it deeper into the ground, creating distinguishable soil horizons. Two thousand years later, the soil is thirty feet deep, it erodes less than one millimeter per century, and the redwood rises 350 feet into the air. We picnic beneath it, leaving an apple core and some cheese rinds as offerings.

Yet the matter that has accomplished this miracle has been largely recycled. If all the litter that had fallen on the forest floor in that time had remained there, the tree would be standing sixty feet deep in the stuff. Instead, the mix of acid humus and still-integral fallen needles, cones, and twigs is a mere four inches thick, and has never been thicker. Its steady decay has fed the tree.

To clear-cut the forest is to break the millennial rhythm of events. Soil destruction is rapid and can be catastrophic. The alkaline ash of burned-over trunks masks these effects for a few years, since it amounts to an all-at-once shot of plant nutrients, but soon it produces accelerating deterioration. The alkali makes silt and clays disperse, shattering the soil's structure and forming a tight crust. Falling water runs off the surface, carrying away nutrients with it. Furthermore, the soil, exposed to the rays of the sun, quickly becomes hotter and digests the leftover organic matter, whose growth-promoting nitrogen soon bleeds through the soil and disappears.

But the most desirable virgin soils are not from the forest. In a forest, at least two thirds of the organic matter resides above the level of the soil. In a prairie, the proportion is reversed. The great prairie soils of the American Midwest and of the Russian steppes are rich with the remains of millennia of dense, sinuous roots that have lived and died in a soft mineral accretion formed from windblown silts high in calcium and other bases. Not only has the root mass given its organic residue for fertilizer; its sugars have also attracted a vast microflora, and its polysaccharides have glued smaller grains of soil into larger aggregates that permit easy passage of food-bearing water and air.

In the 1930s, Hans Jenny studied virgin prairie soils side by side with cultivated soils in Missouri. He found that after only sixty years of cultivation, with zero erosion, the farm soils had lost one third of their organic matter. As a result, there was no longer the same level of glue to bind the soil into aggregates, and instead of rising like yeasted dough, it was collapsing into heavy slabs.

Only by replacing what you take can you keep a soil fertile. As geneticist Wes Jackson suggests, nature must be the standard. To be responsible to the soil is to respond to its gifts with our own.

II

The Matrix

FIRE AND ICE

When I was about four, I used to love to watch oatmeal boil. Inevitably, my mother would throw it on the stove and rush off to make bag lunches. I sat on the red step stool, waiting for the moment when the bubbles first appeared.

A watched pot may never boil, but a watched *oatmeal* pot positively roils. After a stray bubble or two, the stuff would begin to well up in the middle, sending out ripples that reached the pot edge, then dived under the surface again. Suddenly, a whole new mass of light tan bubbles would appear. They hissed as they multiplied, climbing quickly to the edge of the pot and boiling over. Inevitably, my mother would come running, and I would clap my hands in delight.

When the Earth was still new, the surface was a liquid mass, boiling like oatmeal in a pot; the liquid rose to the surface, then moved across it and folded under again. The silica in the melt was lighter than the other elements, so more silica-rich mass would push up and remain. As the silicon cooled, it formed the expanding nipple of a proto-continent.

The silica was a float or a bubble, in whose matrix also came the elements that would make life possible, especially calcium, magnesium, phosphorus, and potassium. At the beginning, this was an ordering process, where liquid matter rose in order to become a resting, structured solid. Once the crust of the earth was formed, however, the order-making convection currents suddenly became intruders. Where they welled up, they no longer had free access. Instead, they had to punch through crust, spewing their half-melted magma over an existing landscape.

This is how volcanoes have behaved to this day. Ultimately, all minerals on the surface of the earth were derived from this process, and fully two thirds of the current crust was thrown up by volcanoes in the last two hundred million years. When the magma is richer in basalt and the other heavier minerals, it may extrude quietly, building a mounded shield volcano, or simply flooding out of the ocean-bottom cracks where tectonic plates diverge. The magma richer in silica is more light and viscous; it bubbles up underground, sometimes forming great batholiths that, when eroded, become mountain ranges like the Sierra Nevada and the Andes. In fact, the name given a volcano made from this sort of magma is called an "andesitic" volcano.

The andesitic volcanoes, many of them located along the Pacific "ring of fire," the system of tectonic plate boundaries that encircles the North Pacific, produce spectacular destruction. Almost exactly a century before the Mount St. Helens eruption, Krakatoa Island in the South Pacific exploded. Its four eruptions created explosions that were heard in Australia, more than two thousand miles away. The volcano literally blew the island to pieces, making a three-hundred-foot-deep trench where a six-hundred-foot elevation had been. The ash thrown into the stratosphere circled the Earth for years, bringing fresh minerals to soils around the globe.

The lava, ash, and dust renew the surface of the Earth with the necessary mineral elements that over eons are carried down to the sea.

Mount Vesuvius erupted in A.D. 79, killing twenty thousand people in Pompeii and Herculaneum, among them Pliny the Elder, the great natural historian. He died with what must have been the most spectacular observation of his long life, considerably less believable than, for example, his thesis that buzzards were impregnated by the wind. The land was black, abandoned. But more than a millennium and a half later, in A.D. 1611, a traveler observed the following scene at Vesuvius:

> *The throat of the volcano at the bottom of the crater is almost choked with broken rocks and trees that are fallen thereon. Next to this, the matter thrown up is ruddy, light and soft; more removed, blacke and ponderous; the uttermost brow that declineth like the seats in a theatre, flourishing with trees, and excellent pasturage. The midst of the hill is shaded with chestnut trees and others, bearing sundry fruits.*

Vesuvius was blooming.

Not all magmas are equally fertile. Those that come from andesitic volcanoes generally yield granites and other acid igneous rocks. Those from basaltic volcanoes yield basic igneous rocks. The former, shiny with insoluble quartz and light in color, tend to make sandier and thinner soils. The dark minerals in granites, the potassium feldspars in particular, give these soils their fertility. The basic igneous rocks, on the other hand, are rich in black biotite, hornblende, and augite, and the greenish olivine. In the salt-and-pepper of a chunk of basalt, there is a good deal more pepper than salt, and these dark minerals are rich in the bases that make a fertile soil.

So much for fire. What about ice?

The soils of the early Earth were cut by gradual weathering, not by ice. Only our recent soils are ice-made, but perhaps without them, the human expansion over the Earth would not have been possible. In the time of dinosaurs, the weathering of the igneous rocks coupled with the decay of organic life and the biosynthesis and laying down of limestones by diatoms, corals, and other creatures would have been quite sufficient to make fertile soils. In fact, the lush fern and cycad forests that the dinosaurs inhabited might never have seen an Ice Age. Though it is possible that the ice has come before—it probably did come around two hundred million years ago—the Ice Age that cut the cloth for most of our present soils was an anomaly.

The fertile soils of the prairies, not only in North America but also in Russia and China and wherever they occur—the black mollisols and chernozems—have fueled the industrial civilizations of the modern age. They are very, very young, the most recent advance of glacial ice having turned back only fifteen thousand years ago. As the ice retreated, it left glacier-ground dust in the outwash plains. The rising winds lifted and spun these dusts out over vast areas of the flattish stable platforms of the world. This fertile dust, lying in deposits up to a thousand feet deep, is really no more than a stirring in the ashes of the earth. From it has sprung the flame that we refer to as civilization.

The Continental glaciers never extended over the entire present-day black-soil belts. Indeed, the soils are richer where the glaciers did not quite reach, because the fertile silt falling there did not fall atop the grittier matter, the rough tills and drifts of dropped boulders, rocks, and pebbles.

A glacier becomes a tool for cutting and polishing first by picking up stone from the bedrock. The part of the frozen surface in contact with the ground sends water into cracks in the bedrock, which on freezing breaks out hunks of stone. These become grit of the glacier's sandpaper, adhering to the bottom of the ice and scraping along carving grooves and grinding smaller or weaker rocks into powder. The powder advances with the glacier front or flows slowly out suspended in outwash streams. Thousands of years later, as the glacier recedes, the wind lifts the powder and fills the gentle bowl of the craton with the dust.

The fertility of the dust is partly owed to its particle size. Crushed small, it has many more surfaces for chemical reaction than would a large chunk. When a cook cuts up a stick of butter in the saucepan, so that it will melt quicker, the same effect occurs. The fertility is also owed to the nutrients released in the dust—potassium, magnesium, calcium, and others, and to the great deep matrix of silica upon which plants can stand firm and secure.

It is strange to think that indeed this grinding, this plowing in solid rock, is responsible for our existence. A cornbelt soil in the United States is an extraordinary machine, even after half a century of rapine at the hands of industrial agriculture. The soil as a body is continually doing work. An acre of good natural Iowa soil burns carbon at the rate of 1.6 pounds of soft coal per hour. It breathes out twenty-five times as much CO_2 in a day as does a man. Every acre puts out a horsepower's worth of energy every day. Without a soil this productive, we would still be hunting and gathering in small bands.

These soils are not eternal. Far from it. They are young. They will grow old and die. For many years Hans Jenny studied an unglaciated soil at the pygmy forest on the Mendocino Coast; no dust or incoming erosive matter has renewed this soil, and it can support nothing at all but stunted trees and shrubs—plants that grow nowhere else—as well as lichens. Because they are porous and because water carries away their nutrients, soils eventually deteriorate.

But we can make them run longer. One motive for protecting the soil is the certainty that it is fragile. It does not have the same unchanging character as a mountain or a river; it is a recent and ephemeral product. We owe it our lives and our energy, and the bodies we give back to it are not payment enough.

DIRT

"DIRT" IS A GOOD WORD. It goes straight back to the Anglo-Saxon and the Old Norse. Like "love," "fuck," "house," "hearth," "earth," "sky," "wrath," and "word," it is short, strong, and leaves a taste in the mouth. Therefore, even before you know what it means, you want to get ahold of it and chew it.

Many people would rather use the word "soil." I met an ecologist in a parking lot one day, getting out of his car. He asked me what my book was about. "Dirt," I said. The man scowled. "Soil, you mean," he corrected.

"No, no. I mean dirt," I insisted. "The stuff kids play in, the kind of road that begins where the pavement ends. Dirt."

An Englishwoman said, "But *dirt,* well 'dirt' among us English is the word for . . . excrement. You know, as in 'dog dirt.' If you are not going to use 'soil,' then for goodness' sake, use 'earth.' It's more spiritual."

"When I say dirt, I mean dirt," I replied. "Earth" can be confusing, because to me it means the whole ball of wax. "Soil" sometimes strikes my ear as sexless and ugly. It makes the mouth taste of sour old nurses who complain, "Mr. A. has *soiled* himself again!"

It takes dirt to grow an oak from an acorn. It takes the rot and the shit that is the root meaning of "dirt"—*dritten* means "shit" in Old Norse. It takes the hot and the wet to awaken the cool order of the mineral world.

Turds no less than rocks and roses are repositories for the energy of the sun. Dirt is where those three meet and meld, to transform the surface of the world and the air that we breathe.

Even into this century, when a country girl was going to be married in France, they fixed the amount of her dowry according to the weight of the manure produced on her father's farm. And until quite recently, if you sold a farm, you always got a credit for the amount of compost that you'd saved. That is what I mean by dirt, the stuff of husbandry.

I mean the stuff that my father used to crumble in his hands and say softly, "That's good black dirt, that is."

THE THEORY AND PRACTICE OF MANURING

Muck is the mother of the mealbag.

—TRADITIONAL IRISH SAYING

A GREAT DEAL OF THE WORLD'S WISDOM is contained in manure. Not only the grain in the mealbag but the full-blown rose are, in one sense, the gift of turds.

To be accurate, manure isn't just shit. It is *both* the dung and urine of an animal, the latter often contained in soaked straw or other bedding. It's important to grasp this dual nature of muck, because while the solid feces are comparatively rich in phosphorus, they contain only about one third of the manure's nitrogen and one fifth of its potash. The greater proportion of these two important nutrients is contained in the urine.

A whole web of organisms in the soil eats manure, cleaving the organic molecules into simpler ones, using some of the results to feed itself, pushing some back into the soil where something else munches it. Then the flatworms and the mites and the beetles and the springtails chase the fattened bacteria, fungi, and earthworms, devouring them, building their own cell walls, excreting the rest. At the death of insects, the chitin-digesting actinomycetes go to work on their exoskeletons, cleaving the tough shells into food and releasing that unmistakable odor that Pliny called "divine," sweeter than any perfume and the only criterion by which to judge healthy soil.

Viewed from this perspective, the process of manure making is slightly comical, yet undeniably attractive in its variety and efficiency. From closer up, it is messy. The cow, horse, chicken, sheep, dog leaves its pie, manure, dung, droppings, dirt, to the tune of roughly two billion tons each year, enough for a three-foot layer over all the home gardens in America.

Who wants to pick it up? Less than a quarter of the manure (including both feces and urine) that these animals drop is usefully returned to the soil. Take the case of New York City. "Sanitation takes it away," says P. O. Oliver of the New York City Mounted Unit. "It's piled out back there," says Bob of the Wichita Zoo. "We pay a guy to cart it off," says the owner of a horse stable not three city blocks away from a community garden that is starving for rich soil. And all over America, urban people wrap their dog's doo fastidiously in old newspaper and chuck it in the trash.

If it's any consolation, modernity and the flush toilet are not entirely to blame. Though among the colonists who destroyed New England's already thin soils in less than a century were some few who had the sense to follow the first-century Roman writer Columella's recommendations and return the manure to the cultivated soil—in some towns, William Cronon reports in *Changes in the Land,* there was a weekly lottery for the right to have the town sheep spend the night on one's land—the majority let their stock roam free, diluting the benefit of their droppings over acres of pasture and forest.

I propose a new national symbol: not Smoky the Bear or an eagle but a colonist planting an apple tree over the old outhouse hole.

JOHN ADAMS'S MANURE PILES

JOHN ADAMS WAS SECOND PRESIDENT OF THE UNITED STATES, the great friend of Thomas Jefferson, and the proudest manure man in colonial America. He was so sanguine about his own compost heaps that his grand-nephew and editor, Charles Francis Adams, excised many of the manure-related entries from John's journals. This was wrong because he, among all the gentlemen of his time, understood the ground he walked upon. Here is some of what was excised:

JULY 1763, FROM THE DRAFT OF THE "ESSAY ON AGRICULTURE"

In making experiments, upon the variety of soils, and Manures, Grains and Grasses, Trees, and Bushes, and in your Enquiries in the Course and operation of Nature in the Production of these, you will find as much Employment for your Ingenuity, and as high a Gratification to a good Taste, as in any Business of Amusement you can chuse to pursue. The finest productions of the Poet or the Painter, the statuary or the Architect, when they stand in Competition with the great and beautiful operations of Nature, in the Animal and Vegetable World, must be prounounced mean and despicable baubles.

BRAINTREE, MASSACHUSETTS. 25 JUNE 1771, "RECIPE TO MAKE MANURE"

Take the Soil and Mud, which you cutt up and throw out when you dig Ditches in a Salt Marsh, and put 20 Loads of it in a heap. Then take 20 Loads of common Soil or mould of Upland, and Add to the other. Then to the whole add 20 Loads of Dung, and lay the whole in a Heap, and let it lay 3 months, then take your Spades And begin at one End of the Heap, and dig it up and throw it into another Heap, there let it lie, till the Winter when the Ground is frozen, and then cart it on, to your English Grass Land.—Ten or 20 Loads to an Acre, as you choose.

BRAINTREE, MASSACHUSETTS. 8 AUGUST 1771

Have loitered at home most of the past Week, gazing at my Workmen. I set 'em upon one Exploit, that pleases me much. I proposed ploughing up the Ground in the Street along my Stone Wall opposite to Mr. Jos. Fields, and carting the Mould into my Cow Yard. A few Scruples, and Difficulties were started but these were got over—and Plough, Cart, Boards, Shovells, Hoes, &c were collected. We found it easyly ploughed by one Yoke of Oxen, very easy to shovel into the Cart, and very easily spread in the Yard. It was broke entire to Pieces, and crumbled like dry Snow or indian meal in the Cow Yard. It is a Mixture of Sand, of Clay, and of the Dung of Horses, neat Cattle, Sheep, Hogs, Geese &c washed down the whole length of Pens hill by the Rains. It has been a Century a Washing down, and is probably deep. We carted in 8 Loads in a Part of an Afternoon with 3 Hands, besides ploughing it up, and 8 Loads more the next forenoon, with 2 Hands. I must plough up a long ditch the whole length of my Wall from N. Belchers to my House, and cart in the Contents. I must plough up the whole Balk from my Gate to Mr. Fields Corner, and cart in the Sward. I must enlarge my Yard and plough up what I take in, and lay on that Sward; I must dig a Ditch in my fresh Meadow from N. Belchers Wall down to my Pond, and Cart the Contents into my Yard. I must open and enlarge four Ditches from the Street down to Deacon Belchers Meadow, and cart in the Contents. I must also bring in 20 Loads of Sea Weed, i.e., Eel Grass, and 20 Loads of Marsh Mud, and what dead ashes I can get from the Potash Works and what Dung I can get from Boston, and what Rock Weed from Nat. Belcher or else where. All this together with what will be made in the Barn and Yard, by my Horses, Oxen, Cows, Hogs, &c, and by the Weeds, that will be carried in from the Gardens, and the Wash and Trash from the House, in the Course of a Year would make a great Quantity of Choice manure.

London. 8 July 1786

In one of my common Walks, along the Edgeware Road, there are fine Meadows belonging to a noted Cow keeper. These Plotts are plentifully manured. There are on the Side of the Way, several heaps of Manure, an hundred Loads perhaps in each heap. I have carefully examined them and find them composed of Straw, and dung from the Stables and Streets of London, mud, Clay, or Marl, dug out of the Ditch, along the Hedge, and Turf, Sward cutt up, with Spades, hoes, and shovels in the Road. . . . This may be good manure, but it is not equal to mine. . . .

THE COMPOST MAN

THE BRAND-NEW ORLANDO, FLORIDA, AIRPORT appears to be made exclusively of plastic: fuchsia plastic, aquamarine plastic, tropigreen plastic. One suspects that even the glass, the toilet bowls, and the candy bars are plastic, and everything everywhere is covered with advertising logos. I am looking for a man named Clark Gregory, who is supposed to squire me all over central Florida in order to show me compost.

This is not the natural habitat of a compost guru. In fact, if ever a structure were made to resist composting, it is this airport. I fear that I may miss him. Is he that smiling bearded man in the checked shirt? He looks all-natural enough. Nope. The guy with the gut? He eats well, he could know about digestion. No, sir. Well, I guess he didn't show.

I should have known better. At the base of the escalator in the baggage claim area stands a tall, gawky Southerner in a billed cap and bright green T-shirt that reads, CLARK GREGORY, COMPOST MAN.

Johnny Appleseed walked around with his cookpot on his head; Gregory travels by plane and rental car. But both of them aim to change the landscape. Sure, an apple tree is sweeter-smelling in bloom than a steaming pile of freshly mixed crab scraps and pine bark, but both vivify the earth. In literal fact, the apple is the gift of the rot that takes place all around it, so Gregory, a.k.a. "the compost man," and John Chapman, a.k.a. "Johnny Appleseed," are brothers.

When we settle into the white generic midsize car, Gregory heaves a weighty freezerbag of black, twiggy stuff into my lap.

"Smell that!" he exclaims.

What is it?

"Scallop viscera compost. The stuff we're on our way to see in the making."

As I gingerly unzip the lock, I wonder what is the olfactory equivalent of being blinded.

To my surprise, the odor is sweet and earthy, with an orangy tinge at the edges. It's like herbal tea mixed with fine topsoil.

Out of the corner of his eye, Gregory watches my eyebrows rise and smiles. "Ninety-six tons of scallop viscera, twelve hundred yards of shredded pine bark from a log builder, twenty-four tons of orange peels, and nine tons of shredded water hyacinth," he intones.

"What?"

"That's what it's made of," he says.

Gregory drives on, his big bony knees jutting out on either side of the steering column.

"It doesn't stink," I say, not quite believing my own nose.

"Not hardly."

Among Gregory's earliest memories is one of his mother burying fish heads in the garden soil of their Daytona Beach home. Later, he tried to study composting, but his Hungarian-born professor sent him to Europe, because there was hardly a world-class composter in all North America.

For three months of 1972, the young man slept on the floors of homes, dorms, and barns throughout England, Germany, Switzerland, Czechoslovakia, and Austria, watching how they did it. He saw composting in vessels, in bins, source-separated and jumbled together, municipal and agricultural. In Zurich, he watched the citizens compost their apartment trash in bins provided by the city.

Part *wanderjahr*, part vision quest, Gregory's trip gave him an evangelical fervor that has never left him. For twenty years now, he has wandered America, preaching the gospel of compost. He tells cities to separate their trash and compost the organic fraction, he tells governors to install compost heaps on the grounds of their mansions, and he tells anyone who will listen that all you need is four construction pallets or a roll of chicken wire to get your compost bin started. "We'll just chip away a little at a time," he says, "and eventually nothing at all will be going into the landfill anymore."

Aren't there things that just have to be thrown away, I ask.

"There's no such place as 'away,' " he replies.

"So all of those wastes from the farm, the home, the lumberyard, and the fishing boats shouldn't be going to the landfill?"

"It's not waste," says Gregory. "It's not waste until it's wasted."

He seems to have an aphorism for every situation. As the little white car scoots west into rural Florida, away from sunbelt splendor, I am his captive and willing audience.

In Georgia, where he comes from, Gregory is consultant to an enlightened few chicken ranchers who don't do what the rest do: discard two million tons of chicken litter per year and one thousand tons of dead chickens per week. A single ranch contains more than seventy thousand full-sized hens, each confined in a small cage. For two years, pumped full of hormones and drugs, the chickens lay and lay. In many cases, they die, because their egg-laying organs blow right out their tails. On a big farm, a dozen chickens die every day.

Here is what the U.S. Department of Agriculture says to do with them: Mix 400 pounds dead chickens, 600 pounds chicken manure, one 40-pound bale of hay, and 5 gallons of water. Mix in 8×4×4 bin with a rain shelter overhead and a concrete floor beneath. Let stand two months.

Gregory makes a scoffing noise.

"Is that good or bad?" I ask.

"It smells like death," answers Gregory. "They should have said four hundred pounds of hay, not forty, but the farmers are interested in getting rid of dead birds and muck, not using up hay. The trouble is that they think of it as waste disposal, not compost production." And why, he adds, should they put a roof overhead and a floor underneath, when compost is a process that takes place most efficiently in contact with the soil in the wide-open air?

Birds and shit are high in nitrogen, the dry hay high in carbon. A proper ratio between the two makes microbes happy. Provided they have sufficient air and water, the microbes eat, digest, live, and die, raising the temperature of the pile to about 130 degrees Fahrenheit and turning foul decay into sweet life. A billion of them to the gram, they make the humus—that is, the organic part of soil, which is all that compost is.

A dump, a landfill, or a bad compost heap is an example of a failed relationship to nature. Even where people have begun large-scale composting, Gregory complains, they frequently design closed systems that end by stinking up the neighborhood, giving compost a bad name, and ultimately failing. "Never let an engineer near a compost heap," says Gregory. "They take a process that requires air to operate, and what's the first thing they do? They stick it inside a box, so they have to pump in the air at a cost of ten dollars per ton, when they might have had it for free!"

There was a huge building in a Florida city, meant to compost the solid trash from people's homes. To operate it successfully, the builders had devised an elaborate fan system. Whenever the temperature suddenly fell, the moisture-laden air would condense into a malodorous pea-soup fog. The men whose job it was to drive front-loading tractors to turn the piles could not see the scoops five feet in front of them. The plant closed less than a year after it had opened.

"But even if you make a better dead chicken compost," I argue, as we drive past a place called Exotic Acres, "isn't the problem really with the factory farms?"

Gregory slows down and turns left onto a small dirt road. "Here we are," he says. It distresses me that he has not answered my question. Later, I reflect that Johnny Appleseed never tried to stop his neighbors from drinking too much hard cider. He just took their leavings—the seed-filled pulp from the cider mills—and planted it abroad, turning their bad habits into beauty and use.

Composting as Gregory practices it is an act of healing. It restores the right working of a natural process. In that act, the participants are not just functionaries, they are sharers in an act of faith.

He rounds a corner, and brings our car to a halt next to the assembled landfill hierarchy of Brevard County, Florida. Men in suits, sportcoats, and neat field clothes await us. They are pretending to look down the five-hundred-foot-long windrows of composting scallop viscera that stretch away beside them.

But their hearts are not quite in it. Or maybe I am the distracted one. Because as impressive as these windrows are, dwarfing my father's old-time four-foot-square compost bin, they are themselves dwarfed by the mountain that rises behind them.

Sixty feet high, it is the tallest spot in the county.

It is also the dump. Before it was the dump, there was no mountain at all there. The county landfill manager cannot hide the pride in his voice when he tells me that someday this immense plateau of steaming trash—so tall that the full-size graders working atop it look like toys—will be three times higher.

Another manager tells me how, when the government tightened the regulations, they'd had to dig all around and under the mountain, lining it with bentonite clay from Wyoming and with plastic, converting the whole thing into an outsized bathtub.

This raised the subject of leachate, the gross stuff that leaks out of a suppurating landfill—and inspired a third manager to describe how they'd counseled a local junior-high girl on her winning state science fair entry, which demonstrated a working leachate-containment system for the home. Unfortunately, she had neglected one step in the instructions, causing her family to move to a motel for four days while the house was fumigated.

We all laugh. These are bright and witty men. "We know where every drunk and every house of joy around here is," says one. "You can't hide anything. Your garbage tells on you."

It does not occur to them, however, that the mountain behind them is doing just the same thing. They speak dutifully of the benefits that the compost windrows will bring, helping them to achieve the state-mandated thirty percent reduction in inflow into the landfill. But the only one who is really enthusiastic is Ollie King.

King has been hanging around the edge of the group, awaiting instructions. He has a couple days' growth of beard, and wears a worn, blue-and-black-checked shirt and a blue baseball cap. When his bosses call, he takes me up on top of his Scat tractor, cranks it to a gut-shaking roar, and starts to work the rows.

When Ollie is not down here turning the compost with this machine that throws it up behind us like a boat wake, he is up on Mount Garbage, opening up holes to receive the county's latest offerings.

"I like working the compost," he says. "I've been with the landfill three years and four months. Down here, it smells for a day and then it stops. Up there on the landfill, it smells all the time."

A whitish cloud of steam rises behind us as we churn up the eight-foot-high rows. He turns neatly at the end of each row and guns the big Scat down the next one. Occasionally, we hit a patch that is less well cooked and a stink of dead meat rises.

Afterward, as we walk down the chocolate-brown rows together, Ollie says of the smell I've mentioned, "That's nothing." He looks around in the heap, combing through the remains of conch, crabs, whelks, and barnacle-covered cans, the wasted "by-catch" of a commercial scallop-dredging operation. He sniffs at a red crab claw that now has the texture of wet cardboard, then discards it. He sniffs a whelk, makes a face, and hands it to me.

"There!" he says simply.

This is not the smell of ammonia or sulphur. It is beyond odor. It makes the gorge rise.

He knocks it out of my hand.

Within another month, however, all this stuff will smell like what Gregory handed to me in the car. Like sweet soil. As Gregory drives me west across the middle of Florida, I no longer see a simple rural landscape. There is no such thing as the country anymore, at least not if by that you mean a bucolic and unspoiled landscape of small farms. Rural America is the place where the cities conduct their most dangerous experiments in mass production and where they seek to dump what they cannot contain. In almost every case, the soil is their victim.

At the same time, in the small towns where communities still survive, you can find the most deadly enemies of the culture of managers:

people who have neighbors and who live close to the soil. They are still capable of common sense.

In neighboring counties, we find two different waste-disposal strategies at work. The poorer county has opened a composting center, where municipal trash is composted, first in closed vessels, and then in windrows. Gregory criticizes the plant because it does not require home owners to pre-separate their trash into compostable and noncompostable. As a result, landfill employees are stuck with trying to separate it all. The process is less efficient, and toxics like leaky batteries are more likely to contaminate the resulting compost. This means that you could use it to revegetate a roadside, but never on agricultural lands.

But this is a trifle compared with the richer county next door. There, the citizens spent millions to build an incinerator that converts all their trash to ash, generating electricity in the process. Our tour guides are polite and well informed, leading us up and down the steel ladders, past the boilers, and into the control rooms that might just as well have been made for a state-of-the-art aircraft carrier.

In the main control room, an operator sits in a big chair like Captain Kirk of the Starship *Enterprise,* directing a huge claw that works on the other side of an immense, hermetically sealed window. Below him is a huge pit full of garbage. In a grotesque parody of those carnival games where you try to drop your claw on a valuable prize, he plunges it into the mounds of trash, lifting them up to the waiting maw of the incinerator.

Were this a fifties sci-fi film, you would suppose that the thing was built by a mad genius and was about to run amok. As it stands, it runs amok just by functioning properly. No contaminants escape through the specially constructed and filtered smokestacks, of course. But what happens to the ash, with its super concentration of heavy metals?

"It is buried in a special containment unit."

Is this the fate of the soil? To become a specialized containment unit for deadly poisons? At least Mount Garbage is visible. You couldn't miss it. But when a buried pouch full of this supercontainment ruptures, you won't know it's there until the ducks start being born with only one wing.

Gregory is oddly silent after our visit to the incinerator. He doesn't praise it. He doesn't criticize it. He won't even talk about compost. We drive on through an eerie landscape of pine forest. All of the trees are small and of about the same girth. Although it is a two-lane road, there are no roadside stands or houses. A smell of hydrogen sulfide hangs in the air.

I ask him what's going on.

"This whole county, pretty much, belongs to Procter & Gamble," he says flatly. "It's a paper forest. The smell comes from the plants where they process the pines into paper."

We do seventy-five miles per hour through this landscape for the better part of an hour and a half. We are riding through the lonely forest on which America wipes its ass and blows its nose.

At last, we pull into the coastal town of Panacea, Florida. There really is such a town, located right in the armpit of the panhandle, making its living on crabs, fish, Jack Rudloe's marine biology supply house, and a bit of modest tourism.

Zelda Barron is the guardian of the Wakulla County Landfill just outside of town. She is puffing a Viceroy, sitting in an overstuffed orange desk chair, at the window of a battered trailer. On a shelf next to her shoulder is a paperback *American Heritage Dictionary,* partly eaten by mice. Large, frank, honest, and not to be disobeyed, she weighs-in trucks, dressed in a red, white, and blue plaid shirt.

Gregory brightens up when he sees her, and the two immediately begin to joke like brother and sister. There are two hundred tons of crab scraps composting together with pine bark and wood chips in her dump, and more is coming in all the time.

A citizen arrives with a load of branches from a tree he's just cut down. She asks Gregory if they can go into the compost.

"Sure," he says, with a wide smile. He is coming back to life.

"Just take 'em over to the shredder there," she says, pointing and forgetting to weigh the incoming truck.

"I've wondered for years why we weren't doing this," she says. "Crab scrap and tree limbs are our biggest wastes. You ought to combine them and make something valuable, instead of just throwing them away."

The landfill is in the second year of a compost demonstration project that Gregory is managing. Already, the demand from gardeners for the finished compost is outstripping the supply.

Gregory takes me out to look at the piles. On the way, we meet a man who manages the dump. I know his name right away, because he is wearing a bright green T-shirt labeled ALBERT HARTSFIELD, COMPOST MAN. I laugh. So does Gregory. Hartsfield's son appears, also in a personalized T-shirt.

Then, I turn around and look into the dump. In the foreground are rough dirt mounds, a pile of shredded tires, and a jumble of old pieces of furniture. But in back is a scene from a Chinese brushpainting.

The long, high mounds of compost are almost black against the dull, rough green of the pines behind them, and wisps of white steam rise from the piles like tendrils of drifting fog. As we walk nearer, the smell of ammonia is strong. Hartsfield volunteers, "I just put fresh crab in. You won't smell a bit of it tomorrow."

I am beyond caring about the smell. The usefulness and the harmony of the project with its landscape are what most impress me. Here in the armpit of Florida, common sense still has power. Later in the day, a citizen of Panacea puts it this way: "If we can drop a bomb down a stovepipe, we ought to be able to deal with our garbage."

That evening, sitting in front of our motel, Gregory is shy and awkward. We will part early in the morning. He presents me with a plastic bag containing something soft. "Here," he says, "this is for you."

I imagine that I have just been presented with a pound of crab compost. But inside is a bright green T-shirt inscribed with the words BILL LOGAN, COMPOST MAN.

THE SOIL OF GRAVES

I WAS IN A MONASTERY IN NORTHERN NEW MEXICO, high in the mountains near Abiquiu. It was February. Behind the chapel there was an open grave, the red soil mounded up beside it. "Has a brother died?" I asked a monk. "No," he answered, "but we cannot dig in winter, so we opened this grave ahead of time, just in case."

An open grave is an open mouth. It disturbs the soil, throwing the wet cold subsoil to the surface. It exhales all the suggestion of the dark. But a grave is also the place where the foul is made fair. It is the way that flesh returns to the generative womb.

The grave seems to interrupt the human story. But the fact is that graves are motherly for the Earth. They wrap up the things of time and deliver them back to the cradle. So that the show goes on. So that nothing will stop the stories from being told.

In this regard, every tomb is empty in the long run. "Putrefaction is the Worke of the Spirits of Bodies," wrote the English Renaissance scientist Francis Bacon, "which euer are Unquiet to Go forth, and Congregate with the Aire, and to enjoy the Sunbeames." Everything wants out. Everything wants to see the sun.

This is not the way it usually seems to us. The folk stories of the grave are full of pale green glows that rise from the tomb, of corpses that sit bolt upright in the grave, of coffins that explode from the evolved gases of decay. The imagery of demons and of hell itself are drawn from the sight of decomposing corpses: lips drawn back, eyes bulging, the skin roiling like an ocean and turning livid pink, black, and green.

But there are other stories that share Bacon's wisdom. In a fairy tale collected by the Grimm brothers, called "Brother Lustig," Saint Peter is called upon to bring a dead princess back to life. To do so, he cuts up her corpse and boils the limbs in a pot, until the flesh falls away, leaving clean bones. Like starched white sheets or a wedding dress, the white bones suggest a readiness to receive once more the complications of the flesh, with all its odors and its staining.

Whitman wondered why diseased corpses, when buried in the ground, did not poison the Earth. "Are they not continually putting distemper'd corpses within you? Is not every continent work'd over and over with sour dead?" he wrote. Yet he concluded in awe at the Earth: "It distills such exquisite winds out of such infused fetor."

Long before Whitman, however, the ninth-century Persian physician Rhazes had intuited this truth. Whenever he sought a site for a new hospital in Baghdad, he brought along a piece of fresh meat. Burying it in the ground, he measured its rate of decomposition, and where the flesh rotted fastest, there he placed the infirmary. It is now thought that the chosen sites were especially high in penicillium, the common soil bacteria from which the drug penicillin is derived. Francis Bacon likewise knew that a shovelful of "churchyard Earth," the soil of graves, would speed putrefaction and regeneration.

While we live, we ourselves are inhabited. A full ten percent of our dry weight is not us, properly speaking, but the assembly of microbes that feed on, in, and with us. Our bodies are the kitchens where our food is cooked, digested, and then burned to cook us. We live until death in a perpetual fever, 98.6 degrees Fahrenheit. When at last we are well done, we begin to cool, becoming food ourselves. More and more ordered, more and more stable, like a good piece of roasted meat, we are made ready. At death, the cornea clouds over, like the eyes of a cold fish, the sign that our first diners are at table.

Decomposition is already under way, though rigor mortis suggests the opposite. The body stiffens, as though to resist decay, but this is simply a sign that the oven has been turned off. One might think that the dead would relax, but for all that exhausted generations have sought rest in the grave, this is not the case. What happens is that the ATP, a phosphorus-rich molecule whose burning provides the muscles with power, runs out and is not replenished. If a person has had a protracted and painful death, the ATP may already be almost gone, so that on dying, they freeze in the precise posture of the last moment of their lives. Dead muscle contracts; it is an effort for living muscle to relax. In order to relax we must burn.

The softening that begins soon after in the corpse has nothing to do with relaxation. Bursting from the thin-walled lysosomes inside each cell, the enzymes that regulated metabolism now metabolize what they had labored to build. They become "autocatalytic," that is, "self-breaking." (This is also what happens when you pound a tough piece of meat to tenderize it. The pounding breaks cells, whose enzymes escape and begin to decay the meat.)

At the same time, the bacterial partners living in the intestines turn from symbionts into parasites, devouring what they had maintained. The capillaries and the tree structures of the blood and lymphatic systems are the roads along which they travel as they spread throughout the body. First come the air-breathing bacteria who exhaust the remaining oxygen in the corpse. They are followed by the masses of anaerobes, which break up proteins, releasing the sulphur that gives the "foul corpse" its cheesy, putrid odor.

The body acquires a bubbling, creeping softness. The metabolism of the bacteria creates gases that swell the belly, making the flesh billow as they migrate along the planes of the tissues. Indeed, in the days when people died at home, you could use this gas to test if a person had departed. The idea was to apply a lighted match to the big toe. Dead or alive, the toe would blister, but if the person was truly gone, the blister would fill with gas and burst.

Over time, the same bubbling and bursting occurs all over the skin as it turns shades of yellow, red, and green. Eyes bulge, the tongue protrudes, and at last the bloated belly ruptures.

All this activity makes the body a more acid environment, propitious for the growth of fungi whose blooms begin to appear on the surface of the skin. Insects and their larvae colonize the remains. Murray G. Motter's gruesome and exhaustive study of 150 corpses exhumed in Washington, D.C., during 1896–97, counted thousands of fly larvae, beetles, true worms, and mites. One large corpse that had been buried for four years was, he noted, "fairly alive with mites, thysanura, beetles and larvae, working on the surface of the cadaver, under clothing."

Unexpected creatures appear in the grave. Springtails are wingless insects that like to live in our lawns, usually at a depth of only a few inches. If they sense a corpse, however, they will readily travel the six feet down to get it. And even in a triple-lined coffin with the inner box of lead, one discovers the proliferating larvae of the clothes moth.

Each grave is like a city beneath the soil, until at last the remains have been reduced to the clean white bones. All the tissues have fed this vast, tiny life that cures its diseases and converts it into rich soil and free air. Even after more than thirty years, Motter could still find intact bones in the grave, though they crumbled to powder at a touch.

The founder of Rhode Island, Roger Williams, and his wife were buried side by side. Later, an apple tree was planted near the grave. When, decades later, the citizens went to find the bodies to rebury them with honor, it was discovered that they had wholly decayed. Not even the bones were left. A nearby apple tree had wound its roots around the corpses, sucking up the phosphorus of the bones and weaving in living roots the shapes of the dead man and wife.

So in the end the tomb is empty, and human forms have been changed into apple forms. The soil of graves is the transformer. It is natural magic. The grave is a memory from which the story of the Earth is told.

But not even the soil of graves is safe from our distempered technologies. Since Egyptian times at least, people have sought to preserve the body against decomposition. The history of food preserving here parallels the history of burial. Salting, pickling, freezing, drying, honey-curing, flaying, bleeding, and eviscerating were all practiced by the earliest Mediterranean civilizations to preserve both meats and the dead. (Our most common practice and the one most responsible for the depletion of the ozone layer, freezing, was pioneered by Francis Bacon, who died of it. He was carried off by a chill caught while collecting snow to cool a chicken.) Sometimes the preservation served ritual ends; sometimes it was for convenience. Alexander the Great was shipped back from

Babylon in a vat of honey. The British naval hero Lord Nelson returned from Trafalgar in a keg of rum.

None of these practices destroyed the soil, though they retarded the work of the decomposers. But Thomas Holmes, known as the father of modern embalming, made his fortune by poisoning the dirt. A Civil War mortician, Holmes ran a thriving business by injecting the bodies of dead Union officers with formaldehyde and shipping them home for burial. Thus, every captain could be his own Admiral Nelson or Alexander, his iconic features preserved for the admiration of the mourners.

Formaldehyde embalming is now common around the world, and is often legally required. The idea is to preserve the living from contagion, but the reverse has been the result. A dead body, messy as it becomes, is not toxic in itself, but formaldehyde is. It coagulates proteins in the corpse and in any creature that tries to consume it. Densely packed churchyards are now so full of formaldehyde that they pollute the surrounding waters. Ignoring the soil's own work, we destroy it and harm ourselves.

THE DUNG BEETLE

THE EGYPTIANS WORSHIPED THE DUNG BEETLE. We ignore him. To the people of the Nile, this beetle, which eats dung, was the scarab, revered not only for its iridescent skeleton and its elegant form, but for the habit we find most disgusting. It buries shit and thus becomes the mother of fertility.

There are more than 120 species of dung beetle. Each one not only eats dung, it collects dung, leaps on dung, and rolls dung into balls so big and smooth-sided that the uninitiated have mistaken them for cannonballs. It then pushes the dung into a private burrow, proof against the flies that otherwise swarm it. In each ball, large or small, the female beetle lays one egg, which she tends carefully, scraping off noxious growths, until her baby is born from the dung's heart.

Most coleopterans, as scientists call beetles, don't give a hoot about their young, and most do more harm than good in the soil. In fact, beetles have a kind of doubly bad reputation among gardeners, because both the larval and adult forms are voracious feeders, often on growing plants of the sort that humans favor. In fact, the larva of a beetle may have a siphon jaw particularly adapted to sucking the vital juice from the roots of a plant, while the adult may evolve a smacking set of mandibles, perfectly formed to devour the same plant's leaves.

The Japanese beetle, that fingernail-sized behemoth whose teeming numbers in the summer attack almost every plant, but especially roses and grape vines, is a case in point. Though it resembles the scarab with its iridescent shell, the Japanese beetle is otherwise a perfect beast. The larvae, nasty little C-shaped white grubs, eat the roots of lawns. The adults multiply at an exponential rate and can skeletonize an unprotected rose garden in an afternoon.

But the Egyptians did not worship the scarab simply because it was more polite than most beetles. They did not worship it because it was cleanly, removing a pile of ox feces in less than an hour. They revered it because it brought the dead to life. It rose out of the shit and took flight on iridescent wings.

The grub of the dung beetle grows and metamorphoses into a young beetle, entirely inside the laboriously prepared ball. Like an egg, to all appearances it is a dead mass. Worse still, however polished it may appear, it is clearly matter that has come out the wrong end of a cow, elephant, monkey, or virtually any other mammal. What the Egyptians noted, however, was that a dung beetle went down into the ground with its ball of shit, and some months later, a new young beetle would appear.

This is why it was the custom to cut out the heart of a dead pharoah and replace the organ with a scarab of lapis lazuli. The scarab was a symbol of eternal life.

Perhaps a more appropriate symbol of renewal has never been found. The scarab renews not only its own young, but also the soil. When you realize that in Africa the record number of scarabs on a single elephant turd is something on the order of sixteen thousand (I want to know who counted them, and how), you get a picture of how many burrows there are and how much dung is incorporated into the soil each year.

The superiority of buried dung to manure on the surface is well known. If the feces are left on the surface, much of the nitrogen contained in them will volatilize and be lost in the air within a few days; if they are buried, less is lost and more is fixed in the soil. Just as important, the buried feces will tend to make more stable humus, improving the soil's moisture-holding capabilities and aeration, and making it a more propitious environment for roots.

As a result, the grasses on which the herbivores feed grow up thicker and faster. The herbivores are healthy and well fed. They defecate copiously. A cow is liable to leave fifteen patties per day; a single elephant turd weighs four pounds. The beetle is the linchpin of this cycle of renewal, which keeps a whole landscape healthy.

The Egyptians admired the facts of the case. We might add that the beetles are uncommonly clever and industrious. When you depend on another's leavings for your meals, you had better stick close to your resource. In areas where large herbivores are common, that does not take much doing, but individual species of dung beetle have found very acrobatic means of keeping themselves abreast of the object of their desire.

In Panama, the scarabs fly up into the treetops every morning where the howler monkeys live. When the mammals let fly, the bugs latch on, falling to the forest floor, where they take their meal, bury it, and return to the treetops for more. Others cling to the hinder parts of their lumbering providers, leaping onto the meal as it drops.

This entrepreneurial spirit pays off because it gives them a jump on the competition. Wherever there is a shortage of dung beetles, there is an excess of flies, since flies are the other sort of insect that by preference lays its eggs in dung. The competition is fierce, and not to the beetles' advantage, because they are slower-moving and subject to being gobbled by birds, monkeys, and the other predators who gather to feast on fresh bug. The quicker a dung beetle gets its find underground, the safer it will be.

We live in an era of delicate sensibilities, and one wonders if the Gilded Age ladies who sported Louis Comfort Tiffany's scarab necklaces—a few made with real dried scarabs—would have done so had they known of the beetle's lifestyle. True, scarabs are somewhat less unpalatable than their near relatives, the burying beetles, who do with dead mice and other small animals just what the scarab does with dung.

Perhaps the burying beetle, not the scarab, is the symbol for our own time. The connection between death and life was a matter of daily knowledge to the Egyptians, but in the "civilized" countries we have pretty well succeeded in exiling the sight of death to the hinterlands of consciousness.

The burying beetle brings it back.

THE ALMOST PERFECT RECIPE

I ONCE ASKED HANS JENNY, the dean of American soil scientists, what made a soil good. He answered with a question: "Good for what?"

Somehow, in my heart I believed he was holding out on me, concealing the recipe that would make anything grow fast and strong. I wanted to know about those proverbial soils where when you drop a seed, you have to jump back to get out of the way of the rising stalk. I did not consider that in the real world such a hot soil might easily result in wheat that grew too much stem too fast, and lodged (that is, fell right over), dropping its seed in the dust.

Neither did I think of those chunky clay soils in the upper Midwest, with a little red in their brown, that break into huge lumps when they are plowed in autumn. A corn seed falling among these chasms would doubtless be lost, I thought, as I teetered among the upturned shards in early October. I did not wait to see the mellowing action of the winter's frosts, breaking the clods and preparing them for the spring fitting that would leave a level and fertile field.

Like every other gardener, I wanted to find the magic soil, the dirt of Eden. The eighteenth-century English agriculturalist Arthur Young called the vale in southern England between Farnham and Alton "the finest ten miles" in England. I wanted to find the finest ten acres in America.

I sent out scouts. All the friends who traveled anywhere took with them a few plastic sandwich bags. I instructed them to scrape the top half inch off any likely-looking plot of ground, get me a bagful of what was underneath, and send it. My eight-year-old son, Sam, and I would do the rest.

Soon we began to receive the sacks: red clays from the Carolinas, heavy black clods from the coastal hills of Northern California; sandy yellow soils from Florida; clay slices from a Wisconsin cornfield; alkaline soils from a Nevada basin; stony gray soils from Vermont; heavy organic bog soils from Minnesota; a whitish surface soil and its rust-colored subsoil from a Southern pine woods; forest molds from New Jersey; red dusts from the rain shadows of the intermountain West; experimental composts made with bark and manure and fish guts.

We set them up in seedling trays, thirty-seven different samples in all, making a patchwork of the soils of America. We applied a half dozen impatiens seeds to each, watered equally, and watched.

Imagine our surprise. Nothing at all came up in the composts. Ever. The California coastal soils, from artichoke fields, gave the seeds a fast start, but the stalks keeled over. The Wisconsin field soils were not spectacular performers, but they got good average germination and produced steady growth. Seeds in a North Carolina red clay sputtered and died. A rogue soil from a Bahamas polyculture, mostly pebbles, did surprisingly well. So did the pine forest's subsoil. But the prize soil of all was from a Wyoming basin. It had been collected by my brother while he was on a motorcycle trip in a barren stretch of country. The soil was really a red dust, so light it would rise into the air if you blew upon it, and it seemed to contain very little organic matter. Nevertheless, when we gave it steady water, we found it outperformed every other soil.

Go figure. The composts that were supposed to work magic didn't work at all. The desert soil bloomed. What was going on here? That was when I finally began to hear what Hans Jenny had been saying, not only to me but in the tenor of his whole work. A soil was not a thing to Jenny. It was a web of relationships that stood in a certain state at a certain time. Only in Eden was a soil eternally fine. East of Eden, a soil would always start and end infertile, but in the maybe 100,000 years in between, it might go through many permutations of fertility, depending on the combinations of climate, mineral matters, organic matter, and slope.

The characteristics of a fertile soil are in the dynamics of that combination. It must not be stiff, dry, or stingy. The Romans liked to speak of the best soils as fat, sweet, and open. To them, a good soil held on tightly and let go lightly. Air, water, and nutrients were all abundant in them, yet they were not held so tightly as to clog up and cease to move. Science cannot really model such a complex dynamic. Virgil and Columella, both of whom recorded their observations of Roman rural matters, insisted that the farmer taste the soil: distill it through a wine strainer with water, and drink the liquor. The best soils had neither salinity nor bitterness, but a sweet and open taste like the smell of fertile soil when it opens in the spring.

A fertile soil transmits forces. Originally, it was the only material on the Earth that could hold water. It therefore became the theater where water, earth, and air could interact, where the earth could express itself in the endless variety of organic life.

Some scientists say that hydroponic gardening—using only water and carefully regulated nutrients—can eventually replace in-ground vegetable growing. It is more efficient and easily controllable. But there are many indications that it could not succeed in the long run. Herbert Koepf and his associates in biodynamic agriculture have experimented over a period of decades with duckweeds and other aquatics grown in strictly mineral media. Though a single generation of the plants might thrive in the mineral medium, the plants did not reproduce well. They lost their vigor and died out. Evidently, only the full soil system provides the proper dynamic.

Neither is organic matter alone a criterion for success, as our experiment abundantly demonstrated. Soils high in organic matter, like bog soils, are notoriously hostile to seeds. What seems to be best is soil that contains four or five percent organic matter and that is well watered and aerated to promote the microbial life that constantly converts the organic matter into humus.

Humus itself is a dynamic creature, comprising a fast-changing part that liberates nitrogen and micronutrients for direct absorption by plant roots, a slow-changing, stable part that holds water, and a porous material that is easy for roots to penetrate.

In the days before synthetic fertilizers, the dynamic potential of humus was so important that there existed a complex formula for calculating its value when property was sold. A person selling a farm would be paid not only for the land itself, but for the value of the prehumic "dressings" he left on the land. He would get credit for the blown-down wood in his forests and hedgerows; for the manure in his stockyards; and even for the plowings that he had done in order to stir and influence the processes in the soil.

The great connoisseurs of the soil have been those who could assess at a glance the character of a piece of land. Probably no one was ever more perceptive in this regard than William Cobbett, the nineteenth-century English author of *Rural Rides.* His descriptions of soil are delicious. Of the typical English soil type he writes: "The short smooth grass has about nine inches of mould under it, and then comes the chalk. . . . The grass never parches upon these downs. The chalk holds the moisture, and the grass is fed by the dews in the hot and dry weather."

This may not be Eden, but to a soil man, it is near enough. Indeed, the better part of what a husbandman knows is what lies beneath the surface of his ground.

III

On Digging Holes

LESSONS IN DIGGING

HANDS WORK BY THE PRINCIPLE OF NEIGHBORLINESS. The competition of one finger against another seldom yields useful results. If you detached the fingers, shook them in a sack, and replaced them in different order, you would no longer have a hand.

Digging and acclamation—the hands delving and the hands uplifted, the hands taking and the hands discarding, the hands scraping and the hands scooping—are the fundamental gestures of the human hand.

Watch children dig. They may sharpen their hands, so the middle finger stands in relief, then scrape two deep parallel stripes in the sand. They may scoop into the sand from two sides, mounding up the beginning of a pyramid between the hands. They pat the side down, then bring shovellike scoops to add to the top. Sooner than a father could wish, they smash the whole structure and begin again, because the point for them is not the building, it's the act of digging itself.

It is perhaps the original choreography of the hands. In African dance, the gestures of planting, of digging, and of lifting up are specifically preserved. Adam made the originals of these gestures when, after he and Eve had been expelled from the Garden, he first delved. "Gravediggers are the ancientest gentlemen," says one of the delvers in Hamlet, "They hold up Adam's profession."

The hand is the symbol of the whole body. It is the only part of us that is made both to give and to receive, both to repel and to grasp. Every tool, every machine made for digging, by child or adult, is a transmutation or a scaling up (or down) of the capabilities of Adam's hand. From the microscopic probe to the kid's stick, the spade and the backhoe, the bucket loader, the grader, the steamshovel, the dragline of mines, the grapple skidder that seizes and uproots mature trees, those novelty cranes that drop like spiders on plush animals, and all the objects that kids turn into digging tools—jar lids, sticks, teaspoons, cups—are stiff, formal, more or less successful imitations of the suppleness of hands.

Perhaps the greatest effort to imitate the hand is the plow. The harrow mimes the hand as it drags its fingers through the soil, loosening and breaking clods. But though harrows were once nothing but a rack of sharpened sticks drawn behind an ox, they are the same in principle today, despite a change to sharper and more durable metal materials. Plows, though they have changed little over four millennia in outward aspect, have sought more and more to hold the curve of the human hand and to imitate its trick of both pulling up and laying down.

Perhaps the finest specimen ever of a plow was made by the eighteenth-century New York inventor Jethro Wood. Wood was obsessed with finding the curve that would lift and turn the soil with the least resistance, making the plow easiest to draw.

He was not alone in this. His sometime correspondent, President Thomas Jefferson, was also in pursuit of "the mould-board of least resistance" and indeed thought that he had found it.

But Jefferson designed on paper, using a grid. Jethro Wood designed on potatoes. People who saw him walking the lanes of his hometown of Scipio, carving away on a spud, soon came to know him as "the whittling Yankee." (Like many western New Yorkers of his generation, he had emigrated from Massachusetts.) His plow was not a product of the Cartesian grid, but was formed directly on a product of the soil.

Again and again, in the letter to the Patent Office of 1819, Wood tries to describe his moldboard, without success. "The figure of the mouldboard . . . is a sort of irregular pentagon, or five-sided plane, though curved and inclined in a peculiar manner," he said. "The peculiar curve has been compared to that of the screw auger; and it has been likened to the prow of a ship," he added, but neither description was accurate. Finally, he gave up trying to describe it in detail: "The mouldboard," he wrote, "which is the result of profound reflection and of numberless experiments, is a sort of plano-curvilinear surface." He then went on to provide a web of measurements so obscure that the document functioned only weakly as a patent, meaning that though his design was almost universally adopted, he saw little revenue as a result.

Even this supple plow, however, is weaker than the hand that modeled it. The machine lacks neighborliness. All the hands or paws of mammals are built on the same idea of digits connected to a stem whose relationship one to the other is extraordinarily various, nearly unpredictable. Where the deer's hoof has two digits, the horse's but one, we have five, thought by scientists to correspond to the original allotment of mammals.

As different as the paws of different animals are, they are all shapely. They speak of a helpful or a fitting relationship of parts. Each finger of the human hand has a psychological and symbolic character, surprisingly constant across different cultures. The little finger is a child, the weak member who must be protected. The second finger waits, it wears the ring, and it follows the high-standing middle finger that so frequently stands for the penis. The index finger is the one that makes point, turns the pages, is the one that Christ raises in teaching. The thumb, in the French Salic Law, was said to be worth half a hand. It is the symbol of human strength and ability. And among all cultures, the palm is a common ground that when exposed means peace and when concealed in a fist means war.

The elements that go to make a hand or a paw are distinguished by the fact that they are both similar and different. Their combination is not a binary question of like/unlike, but a matter of rhythm. Proportions may vary from species to species, but the changes among the digits are always gradual. As the philosopher Alfred North Whitehead wrote, "The essence of rhythm is the fusion of sameness and novelty; so the whole never loses the essential unity of the pattern, while the parts exhibit the contrast arising from the novelty of their detail." To Whitehead, a crystal had too much order to be rhythmic, while a fog had too little. But rhythm is the essence of the structure of a hand.

Likewise, neighborliness and rhythm belong to well-designed tools. Consider the two types of posthole digger. One is a tough-guy male contraption that has two spades that face each other at the end of a long double handle joined with a pivoting part, like a huge pair of tongs. What you are supposed to do is lift it high, then plunge it into the ground, push the two ends of the upper poles apart to trap soil between the spades, then horse the heavy thing out of the ground.

This kind of work is like bad sex. It is in/out, up/down, on/off. Perhaps it will build my pectorals in the long run, but at what expense to my back and my temperament? Even when I had to shut heavy, fast-moving elevators as derrick man on an oil rig, the work was not so mechanical and dull as digging a hole with this implement. In fact, quite the contrary. Though the basic motion was similar—a joining or parting of the hands—the means of producing it was different. I would have to sink to my knees, lean out over space attached to the derrick by a rope, catch the elevators on their way up, shoulder the pipe into their cradle, and slam the doors closed while they kept moving. Come to think of it, I have never since had such a rhythmic job in my life.

The older posthole diggers give this same kind of pleasure. Instead of two straight spades, their business end resembles two hands grasping a basketball, each piece of metal cupped and one set slightly ahead of the other. If I try to describe it further, I will run into Jethro Wood's problem. The important thing is that instead of ramming the dirt, the tool spirals into it. To work it, I operate a horizontal wooden handle fitted crossways atop the long pole as though it were a wheel. The machine translates my pressure and my rotary motion into a force that grasps the soil like a ball. My motions are various, up and down, but also around and around. This is good sex. This is the kind of tool which makes me feel good about telling a friend, "Here, you try it for a while."

Still, it is a pale imitation of the human hand. I wonder if you might have to go the length of ecosystem neighborliness to find something as full of relation and rhythm as the hand. Certainly, the great landscape architect Jens Jensen, who retired from a successful practice to found his School of the Soil on a stony peninsula in eastern Wisconsin, would have agreed. He was fond of calling every creature he could identify a neighbor to every other. In order to make his students better neighbors, he suggested they step on sticks in the woods, to help the wood compost faster, making soils for the maples and birch and cedar.

But we are very afraid of this metaphor of community. To most people, Jensen is a doddering, sentimental old man. Neighborliness in nature! No way. Nature is red in tooth and claw. If I venture to suggest that there is as much cooperation as competition in community life, I will hear all about how stultifying small towns are and how surveys show that young people can't wait to escape them.

Speak of the virtues of neighborliness and manual labor, and people will exhaust themselves describing pernicious smalltown gossip, inbreeding, prejudice, and the history of modern man as a result of the laudable desire to escape back-breaking manual labor. Yet one of the living memories of my life is the first night on a Louisiana oil platform, when Harold Dick, a three-time loser whom nobody ever called Harry, took me aside when I looked ready to fall over. "Hey, Willie," he said, turning me out to look at the distant wellheads flaring in the starry night, "you gonna throw up at least once tonight, you know. But you just keep going back, just keep trying, and pretty soon you'll get the hang of it."

If that is not neighborliness, I don't know what is. It helped him, because it meant he would likely not lose me entirely as a co-worker. It helped the crew for the same reason. And he helped me more than any other "colleague" ever has.

Maybe Jens Jensen was not such a doddering old fool. Indeed, he was a canny enough politician to get more city-side wilderness preserved than any other single individual of his time.

So what if neighbors gossip? So what if manual labor can be demeaning? A real community is born of struggle, shifting loyalties, difficulties, and failures. And the alternative is what we have too much of already: exploitation as a principle, exhaustion, bad sex, a world of insoluble dichotomies in which no idea is ever pure enough, so that we buy ourselves off with rare foods, movies, TVs, cars—anything to take us out of this world.

I want to hold up the hand—anybody's hand—as a sign of value. The principle of neighborliness that it embodies is not a sort of fascism.

We know well enough that every finger of the hand has a use that is separate and that no other finger could do so well. Yet even to do that, the community, the ensemble, is primary. And through the neighborly relation of parts, the hands perform those functions of which prayer is the plainest manifestation: to dig down, to grasp, to lift, and to let go.

ON DIGGING HOLES

HOLES ARE THE ARCHETYPAL PLACE OF DISCOVERY. Who needs to travel thousands of miles to find the new? The most mysterious place on Earth is right beneath our feet.

There are holes for putting in. You put in seed, rootballs, drainage tiles, foundations, treasure, dog turds, the dead.

There are holes for taking out. You take out water, potash, peat, kaolin, potatoes, zinc, oil, the hidden treasures, the ancient dead.

Whatever has come out will eventually go back in again. Whatever has gone in will sooner or later come back out. And I, with my spade or my trowel or my two bare hands, may be the hole's Balboa.

This is one reason that children love holes. The second reason is ants. In the whole memory of the species, how deeply imprinted is the wonder at those puckers in the earth that disgorge red or black or golden lines and swarms of ants? Or sometimes even the furry one that flies up, the ant lion, controlling at once the land, the air, and that mysterious country underground. By the seashore, children experience the same unspeakable pleasure when at certain seasons of the year, every handful of sand they turn over reveals a dozen squirming isopods, tiny armored dinosaurs or living drills who quickly bore down out of sight again.

The third reason is that holes are difficult to dig. Maybe not for isopods, but for us. Holes seem to have a will of their own. When I was seven, my father and I went to plant a shrub—an abutilon with orange-pink paper-thin petals—in the deep acid loam beneath a cypress canopy. He let me jump up and down on the shovel's step—which by the way, is what you are supposed to call the two flat flanges on a spade that you are meant to jump up and down on—as though I were on a sinking pogo stick.

That was the easy part. The hard part was to lever the buried spade head back, grasp the shank low, lift the whole mass with the knees. Mostly, I couldn't budge the full plate alone.

—Could we go play ball now?

—No.

—Is it deep enough yet?

—No. Not yet.

My father used his arm as a measuring stick, comparing the length of the rootball to the depth of the hole. Then, he laid a stake across the surface, to assure that the plant meshed smoothly with the surface of the garden. At the time, I remember, that invention seemed to me to rank with the pyramids and the Colossus of Rhodes. It had never before occurred to me that the surface of the earth was an assembly of holes, each carefully measured to accord in elevation with the others.

We struck a solid rock. My father tilted the shovel in one direction. We hit it again. He tilted it back. Again, the shovel's blade grated on the stone.

At the third strike, the resistance gave way. The spade sank refreshingly deeper.

But at the same time, the hole began to fill with dark water.

Where did it come from? Was it clean? Could you drink it? Had we gotten water from a stone?

My father scowled. He knew that we had split an underground pipe. Of all the luck! I had known his face to go grim and rigid for half an hour after making a mistake like that. But here he had his son lying face-down, laughing at this beautiful, mistaken water, in the presence of a miracle.

So my father laughed too. And in that laughing, we achieved something real out of all the difficulty and the mistakes. We digested that surprise and let it delight us. We lived for a second in the unknown. We did get water from a stone.

You never know what you will find in a hole, so long as you bother to look. Once, I was on a hike in a wild part of New York's Central Park, with my four-year-old son, a ranger, and about a dozen other people. We stepped gingerly over a shallow rivulet of water. Only the kid noticed that this stream was issuing from a hole that came right out of the base of a tree! "Why is that water coming out of the tree?" he asked. Not one other person there had even seen it.

When young John Muir, the Sierra Club founder, went to dig a well on his father's Wisconsin homestead, he reached a depth of more than forty feet at evening. Next morning, his father let him down in a bucket to dig some more. In the cold, still night, the heavier carbon dioxide had sunk to the bottom of the hole. It was all the boy could do to cling to the rope while his father pulled him out of the poisoned air.

When on a winter afternoon my son and I checked a cut bank where we go to watch the soil in rural New Jersey, we found that it was shot through with sheaves of frozen threads of water. The pattern of percolation was perfectly preserved. We could detach big hunks of these, blow out the cold loam from among them, and hold in our hands harps of shining water. (That was when what I'd read in a soil book came home to me: that Earth is the only planet in the solar system where water exists in all three states of matter: solid, liquid, and gas.)

Danger and delight are found together in holes. The first thing you learn in digging is that holes try to undig themselves. Sand slides down around the shovel or your ankles; clods and pebbles rattle down the hole's slope; when a mine ceiling is saturated with water, it becomes heavy enough to collapse. That's the knot of fear: to be buried alive. The staple diet of smalltown newspapers is of daring rescues from caved-in holes. Or like the toad that we kids found stuck in the hexagonal cracks of a dried pond bed. It had gone to sleep in sweet mud and wakened in a closed trap.

But we sought out these dangers, just the way we climbed ropes so high that our organs began to tremble, or swayed in lookouts at the top of a pine where the trunk was not much thicker than our very breakable young necks.

One week, the best fort was an abandoned foundation hole in the oak scrub of a vacant lot. Vines had grown in around it. Bigger kids had pulled big slabs of plywood over for a roof, concealing them under a thick layer of yellow dirt.

No one would ever find us here!

—But what if the big kids come back?

And we made rules, named a club, elected officers, and swore we'd tell no one about this secret hole-fort that smelled of acid and mold.

Then, someone told. The parents notified the town that they had discovered a hazard. End of hole.

It is too bad when people outgrow the love of holes. Sometimes, it happens because the hole has become the symbol of degrading work, of ditchdigging. Or perhaps it is too near a reminder of just how back-breaking work can be. To dig a hole is a glorious thing, and once you are into the rhythm of it, it can be hard to stop. Nevertheless, few adults would start to dig one unless they were told to.

Roughnecks on oil rigs all know, or once knew, the old chant that goes, in part:

"Now gather round, boys, and it's into the ground.
We said, No way, man, we're going to town.
Down on Fourth Street, drinking that wine,
Seeing them whores, what a hell of a time!"

Adam himself did not dig a hole, until the Lord told him to.

But now we have exorcised holes, just as we have exorcised death. Both have become a technical matter, a question of pushing the right buttons. Recently, I saw a folding coffin catafalque on silent wheels. Pop it open, drop the box on, and roll the thing away. When the coffin is gone the catafalque folds up as small as a stroller. Leave a professional to fill up the hole.

—Dead? No, nobody dead around here.

Once, holes were sacred. The Hopi, indeed, conceived of themselves as having emerged from a hole, from the place they had been living as the guests of the ants. (The ants, like Saint Phocas, were good hosts. They fed their guests and tightened their own belts until their waists were as thin as a blade of grass.) All the Pueblo peoples and the Anasazi before them went down into mud-chamber *kivas* for their initiations and spring ceremonials. The pregnant mothers of West Africa dig out the mounds of termites to feed on the calcium-rich earth that the insects have brought up for them.

Nor is the Western tradition without its love of holes. The hill of the Acropolis is honeycombed with shrines and grottoes. All over Europe, the Madonnas have been worshiped in caves and crypts. At St.-Maximin-la-Sainte-Baume in Provence, the crypt of the church is purported to hold the real skull of Mary Magdalene. You have to go down two flights of steps. At the bottom, it smells of piss and wet earth. But to this day, couples come here to scratch graffiti on the walls. Almost always, the graffito is an archway, representing the door to a cave, with a number scratched beside it. The inscription serves as a prayer that tells how many children the couple wants to have.

Most of us ignore holes, or fear them. When there was such a thing as a root cellar, you could at least smell and feel that cold, damp, secret place. Gardeners still sometimes know this love of holes. (How many of you would keep gardening if it were all hydroponics?) For the rest of us, practically the only hole we see outside the bathroom is the refrigerator. There is no denying that it is a very clean and useful hole, but I suspect that refrigerator light is the sort that they have in hell.

PHALLOI AND CALENDARS

THE MOHAWKS THREW A FISH INTO THE HOLE with the seeds of corn. An ancient Scottish farmer would plant a loaf of bread soaked in milk and holy water in the first furrow he plowed. An old black man in the Bedford-Stuyvesant section of Brooklyn will only plant out his sweet potato eyes in a small packing box. Some Chinese farmers moved soil from their mulberry plots into their rice fields and vice versa. And in archaic Yamato, the emperor would engage in intercourse with a virgin in the newly opened furrow.

All these preparations are not merely symbolic. They express the fundamentally sexual nature of agriculture far more directly than does the arcane numerology by which we, each spring, add to our plot three pounds of lime per hundred square feet, five pounds of 5-10-5 fertilizer, half a bale of peat moss, and a sprinkling of bone meal, just for good measure.

The digging sticks of primitive tribes were often made explicitly in the shapes of phalloi, just as are the dibbles sold in every garden catalogue today. (Indeed, I have often wondered whether "dibble" is the diminutive of "dildoe.") And no man who has followed the plow can fail to feel how the thrust of it comes straight from his loins, and how his hips turn as the plow turns.

In fact, the influence is probably the other way around. In the evolution of the angiosperms, the seed-bearing plants, it was demonstrated that the better the contact of the seed with the matrix of the earth, the more likely was germination. So seeds found ways to be inserted into the earth, either by taking projectile shapes, like the wild oats of the California coast; by being buried by animals, as are acorns by squirrels; by supplying their own starter-matrix, as do pomes such as apples and pears, or by being digested in the tracts of animals and excreted with an earthen covering of manure.

Fertility among the mammals is simply a different study in the same vein, one where the male principle, for a change, plays a more precise role. Were we not so uncomfortable with sexual organs, the study of mammalian penises would delight us. In variety of form and fitness for function, they have the beautiful diversity of the diatomaceous world, but with a warmer humor about them. Indeed, it is only through warm blood—expanding into sinus cavities in the veins and arteries of the organ—that it can function as it does.

A human penis is blunt and simple compared to those of other mammals. Of course, the fact that ours are not withdrawn safely into a sheath "when not in use" (as one dictionary delicately puts it) puts a premium on simplicity and smoothness. An opossum, on the other hand, sports a two-petaled penis that looks like a dolphin's open mouth. The anteater's, too, is double, but shaped like a branch pruned back to two stumps. The bull has a strange tower with a deflated balloonlike appendage at the end; the ram has a pile with a slender string attached; and the short-tailed shrew has a long and deeply arched member like a coat hook.

It is rather presumptuous then to say that a digging stick is made to look like a phallus. The phallus is an event in the history of digging.

Yet more sexual than the digging acts themselves are their timing. The basic tool of agriculture is not so much the plow as the calendar. And the calendar, though men have kept it, belongs to women. The calendar ruled agricultural civilizations as much as the clock rules ours. The meaning of time is different for them. When the sage of Ecclesiastes recites his poem on time, "To everything there is a season," he refers to seasons, not to hours.

It is hard from our time-ridden vantage point to imagine the situation in ancient Sumer, when two cities a few miles apart might suddenly find themselves in different months of the year. In their efforts to produce a calendar that would yield reliable information about the times for different agricultural tasks, these newly settled cultures decided on a thirty-day cycle. Unfortunately, it quickly got out of synch—a lunar month actually measures about 27.3 days—so that the month of "barley planting" would occur when it was actually harvest time. Whenever things got out of hand, the local priestly cadre had the authority to intercalate an extra month, to set the schedule back on track. Unfortunately, different cities adopted different solutions, so it would have been possible to miss your business appointments by several months.

This is what happens when men try to deal with women's business. Indeed, the pull of the moon exerts a far more obvious effect upon a woman than upon a man. Menstruation provides an exquisitely sensitive organic calendar.

Until very recent times, not only the rhythm of the moon, but those of the planets and the zodiac were widely assumed to affect the times and types of planting. Astrology formalized these relationships, yet every peasant knew informally how to regulate seedtime by the signs:

Sow peas and beans in the wane of the moon
Who soweth sooner, he sows too soon,
That they with the planet rest and rise
And flourish with bearing most plentiful and wise.

Why should this be? What governs the rising of the sap? What influences the way that the human endocrine system, which, unlike the bloodstream or the breath, is not driven by any internal pump, distributes its fluids? The idea is that the waxing moon draws fluids upward—in the stem and in the bodies of humans and animals—and the waning moon draws them down again. This seemed self-evident to peoples of earlier times. Shakespeare writes of the "tides in the affairs of men that taken at the flood do augur fortune." He is speaking not from metaphor, but from a long tradition of knowledge about lunar and planetary influence.

For planting, this has been taken to mean that plants that fruit aboveground should generally be planted in the moon's second quarter, so that they and their fluids are drawn upward and their fruits grow into juicy ripeness. Root crops, on the other hands, are to be planted in the waning moon, so that their energy is stored underground. Peas and beans and other vining crops were meant to be sown then, too, so that they would establish firm roots before beginning their astonishingly rapid ascent and spread.

THE HOLE TO CHINA

"Mo-ho-ho and a barrel of funds!"

—SUNG BY MEMBERS OF THE AMERICAN GEOLOGICAL SOCIETY

WHEN I WAS SIX OR SO, I still believed that we could dig to China. In particular, Stephanie Buswell and I could do it. We had in fact agreed to do it together. It is true that the previous week we had succeeded in losing a hamster behind the immense cabinets of her parents' garage and were utterly helpless to retrieve it, until we called upon my father. Rigging a paper sack with string and placing grain in the bottom, he lowered it to the terrified rodent, airlifting the beast to safety.

Still, I had no doubt that Stephanie and I were technically capable of reaching China, since the previous week, digging potatoes with my best friend Grant's dad, I had had my first taste of the long-handled shovel. The sandbox would never feel the same again.

We had selected a back corner of the empty lot near school, a suitable place to keep the project private. Our works would be effectively concealed in the tall grass, and few unwitting kids would be likely to fall screaming into the hole, only to emerge head-first in China.

Why did we want to dig to China? Why does every kid want to dig to China? And why was I so positive that when you fell through the Earth feet-first, you would come out the other side head-first? To answer the first question, we were joining the proud company of the second-century scientist Eratosthenes, who on looking into the well near Alexandria when the sun shone into it, succeeded in extrapolating the circumference of the earth. We wanted to live in the presence of wonder and immensity. The second question is easy: in order for a kid to get anywhere without an adult, he or she has three choices: walk, ride a bike, or dig. The shortest, fastest route to China was by digging. As to the third question, it would be absurd to arrive in China standing on one's head. And as six-year-olds, we knew that life was meaningful, not absurd.

The same motives, I think, recently drove grown men to seek to drill to the Moho. The Mohorovicic discontinuity is the boundary between the thinnest layer of the earth, the crust, and the 1,800-mile-thick middle layer, the mantle. It was first noticed in 1909 by the Croatian geophysicist Andrija Mohorovicic as the place where seismic waves, traveling through the earth, suddenly slowed and bent. In fact, it was Professor Mohorovicic's noticing this that first fixed a boundary depth for the earth's crust.

To a lot of people that might have been enough. Indeed, many earth scientists are happy enough to know that the Moho is there and to keep firing off deep explosions to observe the seismic waves and how they travel. Everything they need to know, they reason, can be inferred by studying the way the waves are bent and refracted as they pass through different layers of the Earth's insides.

Other more recent researchers have sought their bliss in the creation of extraordinary core-surpassing pressures at the Earth's surface. Using anvils made with diamonds or with a strange geodesic-shaped molecule called buckminsterfullerine, and heated by lasers, they have simulated pressures that equal or exceed those at the Earth's solid core. This is a pressure on the order of 3,600,000 times the pressure at the surface.

At such a pressure, alchemy, the conversion of one element into another, becomes a commonplace. Certain rare metals have been observed changing into one another in these anvils, and it is thought that with just a touch more squeeze, diamond itself will change into a metal. When the pressure is relieved, however, it changes back. If Stephanie and I had succeeded in our quest, perhaps we would have gone through this electronic looking glass and so really come out of the ground head-first, for proper scientific reasons!

There are people who like to sense the center of the Earth, and people who like to imitate it in their labs. But there still are those who, like Stephanie and me, have to hold a piece of it in their hands. Like the

Bible's Doubting Thomas, we have to feel the flesh of that inner earth, pass through ourselves and come out on the other side. In 1958, a group of scientists casting around for something useful to study in the earth sciences conceived that they would rather hold the Moho's rock in their hands than simply know what it did to waves. Like Stephanie and I, they quickly decided to begin, although they were vaguely aware that the deepest well yet drilled was less than half as deep as the six-mile hole they would need.

Neither were they daunted by the fact that they would have to drill this hole starting in water more than a thousand feet deep in the open ocean. The reason is that the Moho comes closer to the surface there, where the crust is thinner. Where land surfaces and mountains occur, their above-ground weight is surpassed by a bulge of mass underground, so that while the Moho is six miles down in an ocean trench, it is twenty or more miles down on the North American continent. They invented a new technique, called dynamic positioning, to hold a ship in one exact spot on the ocean in any weather for the duration of drilling.

Nevertheless, when push came to shove, they were not much more successful than were Stephanie and I. (And they spent a great deal more money.) The project did succeed in positioning a ship and drilling a few holes to a depth of several hundred feet and in recovering interesting core samples from the ocean sediment and basalt rock.

These basic experiments were the foundation for the deep-drilling ships that would later map the ocean floor, helping to establish the tenets of plate tectonics, but soon the focus shifted to extrapolation, not piercing the Earth's crust.

Stephanie and I, on the other hand, took turns with the shovel for the better part of an autumn afternoon, while the wind bent the brown grass and made a hissing whisper all around us. The work and her blond ponytail made me dizzy. It was hard: the ground would not let us through.

How was I to know that to make this suburb they had scraped off the millennial accumulation of topsoil, in order to render the land flat and tractable, then sprinkled a foot of it back on. Wherever we dug, we hit hardpan, caliche. It was going to be more difficult to reach China than we had thought.

It was not a good year for the essentialists. The Mohole people lost their money. Stephanie and I barely got hip-deep. How were any of us ever going to hold the truth in our hands?

IV

Earth and Stone

EARTHQUAKES AND MOONQUAKES

Mobile being *is the subject of natural philosophy. . . . Nature is a principle of motion and rest in that in which it is.*

—THOMAS AQUINAS,
COMMENTARY ON PHYSICS

THE BALL OF THE EARTH CONSISTS OF A SHELL of maybe a dozen discrete plates that bump, bang, rub, and stretch among themselves, riding atop the viscous liquid crystal surface of the underlying asthenosphere. When two plates converge, they may slide smoothly along each other's margins, but more likely, they will catch and halt. The pressure builds. Then, like a stick bent to the breaking point, they suddenly snap apart, sending out perhaps the hugest and deepest sound in the world. It is so deep that it can't be heard by human ears; it is experienced instead as motion, vibration, displacement of the ground, collapsing buildings, people and objects thrown into the air, and sometimes even as a pulse of light.

On the moon, there are no tectonic plates. Too small and therefore too low in gravity to possess either an atmosphere upon its surface or the heavier metals in its interior, the moon is a cold, slow lump. But deep inside it is the beginning of a distinction between mantle and core, and at that boundary, moonquakes happen. They propagate rapidly, stirring the sterile dust of the surface, which in places is as much as thirty-three feet (ten meters) deep. It is as though the moon were trying to imitate the Earth.

On average, 150,000 earthquakes occur each year on the Earth. Some are deep-focused in the mantle, and imperceptible at the surface.

Others move the soil and the creatures in it with unmatched suddenness and ferocity. When water-saturated sands or clays are put under pressure, the water may force the grains apart, causing solid ground to melt beneath your feet. In 1964, the Good Friday earthquake in coastal Alaska liquefied the sandy soils along the shore, causing hundreds of acres to sink into the sea. If the land is on a slope, it may rush down. The Peruvian earthquake of 1970 sent a landslide down the Andes slopes that buried Yungay, a town of twenty thousand souls, in just a few minutes. In 1971, the San Fernando Valley earthquake in Southern California set off over one thousand landslides in the area. Liquefying fill at the base of an earthen dam was within ten seconds' shaking of causing the dam to fail, which would have killed tens of thousands downstream.

Up to weeks before an earthquake occurs, however, some change happens in the soil and water. The creatures who live there sense that something is amiss. The Chinese were able to avoid the worst consequences of the Hai-Cheng earthquake of 1975 by taking warning from animals that live in the soil. Though it was mid-February, people observed snakes emerging from their burrows and dying on the frozen ground. Rats abandoned their holes to wander the snowy landscape in groups. Prior to other earthquakes, ants have been observed trooping across the soil surface with their eggs held in their mandibles. Rabbits have been seen hopping on the surface, refusing to enter their burrows. Sheep, cattle, and horses have balked at entering their corrals. Fish jump repeatedly, and shrimp crawl onto dry land.

What do they perceive? It may be that they sense a change in the local magnetic fields. Iron oxides in the soil, formed at different depths in different horizons, have their own characteristic magnetic fields, corresponding to their orientation vis-à-vis the poles of the Earth. It may be that when compression and tension deep inside the Earth bend these fields, the animals who live in or on the soil sense the change directly and respond by flight. This may sound far-fetched, but it is equally probable that pelagic fishes, whales, turtles, birds, and other long-distance navigators constitute the world around them and find their way repeatedly to the exact same compass point, by means of a kind of magnetic proprioception. Even the early Polynesian navigators, who found Hawaii unerringly from the Marquesas across a thousand miles of open ocean, were following currents as their clues, perhaps magnetic currents as well as oceanic.

The Tangshan earthquake of 1976, probably the second most destructive earthquake ever, killed 750,000 people. Just before the first shock, which threw adults six feet into the air, many people observed a bright, incandescent, red-and-white light coming from the area of the epicenter. Afterward it was found that some shrubbery was burned only on the side facing this light.

Is it possible that the tremendous, deep vibration given off by the quake from its focus could heat the soil itself to incandescence, so that it emitted light? A witness to the San Fernando Valley quake, whose house was located near the epicenter, observed a steady glow coming from the earth-fill Pacoima Dam, as though it had become an enormous filament.

James Clerk Maxwell conceived of the universe as an electromagnetic realm characterized by different degrees of conductivity and resistance. Long before it was technically feasible to create loudspeakers or color TVs or electric lights, he accurately demonstrated that the properties of electricity and magnetism would give rise to them. Incandescence, he wrote, occurred when a current passing through a slender wire overcame the resistance of the air around it, causing the air itself to become charged. The repeated excitation of the air released light.

Might it not be that an earthquake does something similar on the scale of the Earth? The tectonic plates, moving and readjusting, cause the Earth for a moment to emit light as though it were the sun.

From the Earth's point of view, the whole thing may be nothing more than a moment of peristalsis, a readjustment that helps her to digest her food and to relieve herself. But one can't help wondering what the light is about.

Likewise, the Earth stimulates the moon to shudder and flex, as though she were imparting to her the beginning of a warmer life. A moonquake is caused less by tension between the moon's tiny iron core and its surrounding silicate mantle than by the Earth-moon relationship. The strong, deep-focus moonquakes almost all occur within a few days of the moon's perigee, its nearest approach to the Earth. The Earth stimulates the moon to shudder and flex. Selene turns toward Gaia like a child at the breast.

THE CIRCULATION OF STONE

THERE IS A GLAMOUR to the study of rock. It is so old, so hard, and it has taken the entire force of the planet to produce it. Inside the Earth, heat seeks to randomize motion; outside, the pressure of gravity restrains it. Between these two forces, the whole world is extruded, from the bottom of the sea to the highest mountain. Imagine the two hands between which you even out a mudpie. One pushes down, one pushes up. That's the way it is with the Earth. So what we know as the surface is really only a way station—a point of equilibrium—between opposing forces.

And the surface moves. Very slowly to our eyes, it is true, but one could imagine a being for whom each of our millennia was a second . . . Such a creature would observe the shifting and the collision of continents, the eruption of volcanoes, the appearance and disappearance of mountain ranges, the spread and retreat of ice and whole floras, and the renewal of the seafloor, as clearly as we observe that a living person's chest moves up and down, and she blinks, twitches.

The ultimate fate of the land surface is the same as the seafloor's: it flows, slides, blows, sinks over the edge of the continent and into the suture that returns it to the underground fire. From there, it is once again digested and circulates to a weak spot in the crust, where it is pumped upward again.

This great roiling circulation belongs to the old gods. Nothing smaller than a planet can influence its eonic course. Imagining the undersea mouths where the new crust spills out and where the old crust is swallowed, one thinks of the first goddess, Gaia, she who gave birth to her children only to devour them.

What part does the soil play in this process? It is less than foam on waves. Geologists, who love deep time and the vast forces that make igneous, metamorphic, and sedimentary rock, scarcely recognize soil at all. They call it "regolith," meaning unconsolidated rock, as though it were simply a moment on the way to the cementation of real stones.

But the soil is all of the Earth that is really ours. The seasons, with their heat and their cold, make the soil. The storms make the soil, with water, the most powerful substance on Earth. The winds make the soil, spreading dust across thousands of miles. The tides make the soil, stirring the river deltas and their fertile slimes. And above all, the trees and the plants, the dead and the digested, the eaters and the eaten, make the soil.

The mouth of the Mississippi River spews fifteen tons of sediment per second into the Gulf of Mexico; every American acre gives up an average of more than ten tons of soil each year. Fully seventy-five percent of the Earth's surface is made of such moving or cemented sediment.

Dirt is the gift of each to all. It is not so grand in appearance as even the muddiest shale. An ancient soil may be no more than half a million years old, piddling compared to any rock. In a hot, wet place like the South Seas isle of Krakatoa, new soil may form at a rate of better than one half inch per year, while on dry lands, it may take better than a quarter of a century to do the same; but the metamorphic rock of the shields is more than three billion years old. A tectonic plate of the Earth's crust is as much as fifty miles thick; a soil may be as thin as a lichen. It would seem incommensurable. The soil is so young and fragile compared to stones. But youth and fragility are fruitful.

Soil appears where life does, and its characteristic is to build where erosion destroys. On the face of a stone a lichen takes hold. The lichen digests minerals and is itself digested by the microbes in the air. The combined detritus falls, fills a cleft in the rock. A club moss roots in this compost, lives, and dies. The cleft overflows. Grass seeds blow in, grass grows.

It is the nature of soil to build aggregates: plates or blocks or chunks, full of air and water channels. Gardeners in fact know their soils first and foremost by the size of the particles and the kinds of aggregates they build. Where organic decay and inorganic erosion meet, the conditions exist for a fertile soil, because the two in combination make a tortuous, knotty structure that offers roots the optimum mixture of mineral nutrients, organic nutrients, air, and water.

Everywhere, creatures and minerals together make their characteristic soils. Where the grand circulation exposes different bands of rock in juxtaposition, so the plant communities that come to live on them differ, and the resulting soils do too.

On a field trip once, the great soil scientist Hans Jenny had me drive down a side road in Sonoma County, California. Beside us ran a field of wild grass with oaks scattered on the knolls. Abruptly, the vegetation turned into a scraggly stand of digger pine. We stopped the car to look for the difference. Beneath the oaks, the broken rock was schist; underlying the grassland were sandstones; but the sparse pines grew on a pretty, green stone. Serpentine is the state rock, it's so attractive, but it is also full of chromium and nickel, discouraging to most plants.

Here, we used our eyes to conceive the livingness of the world. I had driven past that landscape all my life, my eyes on the road ahead, noticing the "beauties" but really observing nothing at all. Here, by an oily roadside, with traffic sweeping by, I stepped out of time and into beauty, thanks to a ninety-year-old man who actually knew something. I could feel the Earth spinning on its axis.

We spend our lives hurrying away from the real, as though it were deadly to us. "It must be somewhere up there on the horizon," we think. And all the time it is in the soil, right beneath our feet.

OLD QUARRIES

Man makes an end of darkness
when he pierces to the uttermost depths
the black and lightless rock. . . .
Man attacks its flinty sides,
upturning mountains by their roots,
driving tunnels through the rock,
on the watch for anything precious.
He explores the sources of rivers,
and brings to daylight secrets that were hidden.
But tell me, where does widsom come from?
Where is understanding to be found?

—JOB 28:3, 9–12

I REMEMBER AN OLD QUARRY in the Catskills where we used to swim. The water was so clean, clear, and cold, just the opposite of the muddy and murky lake a few hundred yards to the south.

No inconvenient soil came between the water and the rock. The nineteenth-century quarrymen had made a fifty-foot-high wall. Through a cleft in it, a stream of water now fell. From either side of the waterfall, you could jump, if you dared, and plunge into deep water.

But a quarry never fully belongs to men. It represents a compromise with the old powers of the earth. This quarry was still there, long after its makers and their towns were dust, and after the rains had worn the inscriptions off their nearby tombstones. And I never once dove into that water without the momentary fear that the earth would simply swallow me.

Quarries live by the Earth's time and on its scale.

To get to the quarry sites at the Indiana Limestone Company in Bedford, Indiana, you have to drive through ten acres of stoneyard, where they pile the cut blocks. Stacked three and four stories high, the forty-ton blocks perch roughly atop each other, each marked at the corners with the neat incision marks of wedges. Over here are the pure gray stones, here the blondish ones, and here the ones with veins and mixtures of tan and gray. (All of them are curing, losing the five percent quarry water of the fresh stone in the earth, a process that makes them tougher and more durable.) In a discard pile are stones too pitted with fossils to be valued by the building industry.

Here are the raw materials of whole cities—their hospitals, courthouses, cathedrals, and banks—laid out in enormous open-air supermarket aisles alongside which my little blue four-door sedan looks like a Matchbox toy. Indeed, I have the impression that I have become a character in a child's metropolis, formed of kindergarten blocks, and I wonder where the big bright letters are that should adorn the massive stones.

The quarry is an even stranger sight than the blocks that have come from it. The older holes are filling with a deep-blue water; those that are still active look like the negative space of a city, or like holes left after ziggurats have been extracted foot-first from the earth. In a way, they have. One big hole that we pass—about the size and depth of a small lake—is where the Empire State Building came from. Another was the source of the Cathedral of St. John the Divine.

The quarryman doesn't speak of soil. What we call soil, he calls overburden. And he drags it off the stone with power shovels, the way a father pulls back the covers of his son's bed. In the great limestone quarries of Indiana, you can walk in a swirl and tumble of clayey soil that seems to have been swept together like a crumpled blanket.

The quarryman couldn't care less about the clay. It is too easy to move. He, like his predecessors for five thousand years, must find the way to turn rock into geometry. With his wedges and hammers, he stands on the uncomfortable edge between man and nature. It is little wonder that every ancient quarry is full of religious graffiti: prayers to nymphs and to the Lord God. Alone among mortals, the quarryman can make the earth tremble.

The stones come to shuddering life in quarries. A crack may widen too quickly and bring a side of the mountain down upon you. When you take out blocks of stone, you may find the whole quarry shifts beneath you and a solid block turns into a sheaf of playing cards.

Rock does not sit inertly on the earth. It is already under pressure, both from above and from below. It holds great forces among its folds and twists. Over time, quarries rise into the air, as it were, because as the layers of stone are removed, the force of gravity grows less over the hole, whose floor adjusts upward, like a boat in the water when you lighten it. Sometimes this process releases a twist that has been straining inside the rock for tens of millions of years, and a whole house-sized chunk bursts upward with a sound like a bomb. In some quarries, where they now cut by means of a sword of flame, the rock closes again as soon as the torch has passed, thrust shut by overpowering forces from beneath.

For the better part of five thousand years, quarrymen have been struggling with these processes, using practically the same tools and developing traditions that, along with those of farming, are the most unchanging in the world.

In the sixth century, the fortified Byzantine city of Dara, at the edge of the Mesopotamian plain, challenged the Persians. Procopius, who exhausted himself in praise of Justinian I's part in building it, attributed everything, from the height of the walls to the novel means of supplying it with water, to the emperor's wisdom. In fact, however, the aqueduct was designed by the river itself, which in one terrible storm breached the city walls and flowed through the streets, carrying furniture and pottery with it. At one point, the stream is said to have disappeared—flotsam and all—into a well that Justinian had had dug, and reappeared near another town forty miles away!

The stern magnificence of Dara can still be guessed at today by looking at the fabulously deep and sheer-stepped quarries that line the dirt track leading to the little village that is all that otherwise is left of the great fort. They are reminders that, as Procopius said, at one time the city was "surrounded by two walls, the inner of which is of great size and a truly wonderful thing to look upon (for each tower reaches to a height of one hundred feet and the rest of the wall to sixty)."

The city fell and declined into a village of shepherds, but the quarry became filled with tombs hewn into the rock. So as the living withdrew from the area, the dead replaced them, turning the fantastic negative space of the quarry into their own city.

In 573, the Persians took Dara after a six-month siege. So many died in the last week of fighting that it was said the warriors were like harvesters, "mowing and smiting one another like ears of corn." The last defenders choked the city wells and its river with the dead, hoping to leave at least pestilence among the victorious Persians.

What did the Persians do with the bodies? As the fort's defenders well knew, the Persian religion, Zoroastrianism, forbade bringing dead bodies into contact with the earth, lest they pollute her. So their last gesture was not only a military move, but an intentional sacrilege. To the Persians, the right way to treat the dead was to expose the corpse to the sun, so that the spirit might be drawn to the light.

What did the Persians do with the Christian dead? The archaeologist Oliver Nicholson has suggested that they simply exposed the dead upon the hillside. Perhaps they piled the bodies on the ledges created by the carving out of blocks in the quarry. It would have been the best thing they could do, since in that way the corpses would have been preserved from polluting contact with the soil. The dead, then, with the birds and beasts that fed on their bodies, would have kept watch over the approach to Dara.

Eighteen years later the Persians returned Dara to the Byzantines. As the Christians rode back to their city, they must have seen shining on the quarry ledges the piles of the white bones of thousands of their countrymen. How could they not have been reminded of Ezekiel's vision of the valley of dry bones? Indeed, among the great reliefs carved into the walls of the quarry-cemetery is a picture of this biblical vision.

The prophet wrote: "The hand of Yahweh was laid upon me, and he carried me away by the spirit of Yahweh and set me down in the middle of a valley, a valley full of bones. He made me walk up and down among them. There were vast quantities of bones upon the ground the whole length of the valley; and they were quite dried up. He said to me, son of man, can these bones live?"

The dead of Dara did not rise up and walk, as did the army of Ezekiel's vision. The city decayed to dust. The quarry became a place of fear, where until very recent times visitors were warned of the presence of evil spirits. Perhaps this was because, as Job knows, the dead are stronger. Their city is not erected but carved out of the living rock.

THE FOUNDATIONS OF CATHEDRALS

I HAVE SPENT THE BETTER PART OF THE LAST DECADE at the Cathedral of St. John the Divine in New York City. For most of that time, my office was in the church itself, high up at the Triforium level. Few people in this age have the chance to walk up and down a stone spiral staircase going to and from work. And few have had to gauge the weather outside by looking at the brightness of St. Paul's face in an immense stained glass window.

I love to wander high up along the parapets and beneath the roof on the convex side of the vaults. But more impressive to me than these is the Cathedral's foundation and its well.

Architects love to pore over the plans and elevations of cathedrals. I imagine that if every page ever devoted to arch, vault, and buttress were assembled in one place, you could construct a life-sized model of Chartres out of the paper alone. Yet the number of pages devoted to the foundations of these greatest and heaviest of all Western stone structures would not even fill a briefcase.

This knowledge imbalance has had some comical, and costly, results at my church. When the Episcopalians of New York City decided in 1892 to begin the construction of a massive cathedral for their diocese, they looked at drawings of Gothic and Romanesque cathedrals before settling on a magnificent Byzantine-Romanesque design by the firm of Heins & LaFarge. Of course, the drawings, like those of most books of church architecture, made no reference to the foundations. These were assumed.

The builders began to chip and blast away at the outcrop of Manhattan schist selected for the church's site. It looked like a simple matter, since the outcrop had a very shallow overburden of loose soil. Beneath, it ought to have been solid metamorphic stone. But as Eugène Viollet-le-Duc might have warned them, stone can be more treacherous than clay. Every foot of excavation revealed another web of cracked and twisted rock.

Across the street, a separate team of builders had begun on St. Luke's Hospital. The two teams began as rivals, racing informally to get their holes dug first. Pretty soon, however, the terms of the rivalry changed. Would the cathedral builders hit bedrock by the time the St. Luke's foundation was done? Or would they ever strike bedrock at all?

Angry Episcopal missives flew back and forth; there was talk of abandoning the hole. Then, the industrialist J. P. Morgan wrote a blank check, saying "Dig it and be done." One half million dollars (twenty million dollars in today's currency) went into that hole, and St. Luke's Hospital was wholly built and dedicated, before the church's foundation was done.

Forty million years previously, the diggers would have found no difficulty, since at that time the schist had not yet been twisted and fractured into the jigsaw puzzle of fragments that it is today. On the other hand, the Hudson River had not yet appeared, either, so Manhattan real estate would not have been so valuable.

It is a good thing, however, that they did find bedrock. Though Heins & LaFarge were fired and the church continued as a massive Gothic cathedral under the direction of Ralph Adams Cram, it was the weightiest Gothic undertaking ever. With its towers still incomplete, the cathedral weighs about 253,000 tons.

Bedrock, as solid as it sounds, is an abstraction that refers to stone that has not yet been deformed. True bedrock can carry about a thousand tons to the square yard in load, but finding it can be difficult. Regardless, it is perilous to forget that buildings don't simply sit on the ground with the weightlessness of pen lines on paper; they rest on the surface of the Earth.

The great foundations of history have all been for ceremonial buildings, and the early Mediterranean experience with them would have made very vivid the parable in the Bible of the well-built house:

Whosoever heareth these sayings of mine, and doeth them, I will liken him unto a wise man, which built his house upon a rock: And the rain descended, and the floods came, and the winds blew and beat upon that house; and it fell not: for it was founded upon a rock. And every one that heareth these sayings of mine and doeth them not, shall be likened unto a foolish man, which built his house upon the sand; And the rain descended, and the floods came, and the winds blew, and beat upon that house; and it fell; and great was the fall of it. (Matthew 7:24–27)

In fact, even the solidest foundations in practice might not hold, since in addition to salting fields, a conqueror who wished utterly to destroy a city would rip out its foundations. "I have destroyed the city right down to its foundations," said King Sennacherib of Nineveh, referring to his conquest of Babylon. "I have followed this with a flood, and I have ordered the materials extracted from the lower foundations to be thrown into the Euphrates that they may be carried to the sea."

In the ancient Near East, furthermore, many foundations sank of their own accord. The cities of the Mesopotamian plain were founded on spongy alluvial soils, and the first great ziggurats demonstrated the principle of isostatic adjustment admirably: every few years, the priests would have them built a few steps higher to compensate for the sinking of the bottom story into the soil.

It is a strange fact, but what builders began to learn at the beginning of the Monumental Era—perhaps a good name for the Western world—was that the surface of the Earth was itself a kind of sea, requiring piers and rafts to support the brick and stone towers raised upon it. The first thing the ziggurat builders learned, indeed, was to plait layers of woven reed mats between the courses of bricks, which made a simple, slender raft to spread and equalize the downward force of the temple and the spreading splash-response of the soil. This is the same principle used by the Greeks when they joined stone slabs with iron pegs to found their temples and when the Romans and we moderns followed with concrete rafts on which to found our structures.

But perhaps the greatest foundations of all are those of a few of the Gothic cathedrals, particularly Amiens. A dozen courses of stone and mortared rubble beneath ground level formed a raft for the entire structure, so that where the piers of the nave and choir descend into the earth, a closely integrated pyramid of underground stones spreads into the ground until it rests on the virgin clay more than twenty feet beneath. To see the full drawing of this structure from the top of the flying buttresses down to the foundation is to comprehend the true structure of the Gothic cathedral. Each pier has the form of an arrow, with the foundation stones playing the part of the feathers that keep the point on target. The drawings look like a quiver of arrows, poised for a shot at the divine.

Few structures of any sort have ever had such an honest relation to the ground they stand in. But the cathedrals needed to be honest buildings, because by their very weight they intruded into the earth, insisting on bringing it into relation with the heavens.

Often, cathedrals were built with an underground crypt, a damp place where relics were stored, and where a spring might rise. Even at Saint John the Divine, the weight of the building and the need for a deep foundation caused a living spring to rise in the crypt. The base of the Cathedral is now beneath the water table. But in at least two cases, cathedrals were intentionally built to incorporate the forces of the Earth, in the form of a venerated water source.

Legend has long had it that the Cathedral of Notre Dame de Chartres was built on the site of a druid's cave, where pagans had worshiped a dark-skinned virgin mother goddess beside a hidden well. It was said to have been an important place of pilgrimage long before the birth of Christ.

In the mid-seventeenth century the pious Vincent Sablon spoke of the well and its dark-skinned Madonna in great detail, recounting at least the well as a present fact. Known as the Well of the Strong Saints (Saints Forts), it was said to have taken its name from a persecution of the first century, when the Roman emperor had had numberless Christians killed and stuffed into the well. Far from stopping the cult of the Virgin, however, the emperor's action simply provided the nascent church with more martyrs to venerate.

According to Sablon, the generations of early Christian pilgrims who flocked here worshiped their black Madonna as Notre-Dame-Sous-Terre, or Our Lady Under Ground. At the well, they took water both for baptisms and for cures. Even after Bishop Fulbert had built his cathedral here in the ninth century, the faithful still descended into its dark crypt to visit this most sacred place.

But by the end of the nineteenth century, the whole story smacked of legend. Where was this famous well, anyway? There was no sign of it in the musty crypt of the twelfth-century cathedral, and indeed there stood a wall where it was supposed to have been. In any case, it held no place of honor, even were it really there. It was too messy and too wet an idea for the rational religion of the turn of the century.

Then, in 1901, the archaeologist Merlet found it. The wall that covered it was, in fact, a late seventeenth-century addition to the church. Prior to that time, a gap had been left in the foundation courses, specifically to leave the well revealed.

Merlet found it to be a rough tank about nine feet square, carved in the limestone of the hill. When he scooped out the clay with which it had been filled, he found that the well was a hundred feet deep. The well had been partly filled with earth as long ago as the twelfth century, when its great depth might have endangered the deep foundation stones of two pillars of the Gothic choir.

The high-rising spires of Chartres rest on this foundation of dark water. Though all the legends of druidical rites may be fallacious—they do not seem to predate the fourteenth century—the point of the legends is a good one. It is unmistakably true that Chartres was intentionally built over an ancient well, not so important that it could change the orientation or placement of the Gothic cathedral, but important enough that the church would preserve a place for it, even when logically it should have been covered with a wall.

GROUNDWATER

I ONCE HAD THE PRIVILEGE OF LIVING WITH A WELL. It was a civilizing influence. My wife was six months pregnant with our first child; we stayed in a farmhouse at the very top of a village in Spain. The stone-built well, an *aljibe,* was right outside the front door, on a flagstone patio shaded with a grape arbor. To the right of the door, a large window opened onto the kitchen. The broad shelf of this window was the way that water got into and out of the house. The cooking water went in through it; and we washed dishes in a pan directly on it.

Because we had no bathroom, we bathed in a large tub that we would fill with well water and set in the sun to warm. Alternatively, we used our host's ingenious contrivance of a large bucket fitted with a shower spigot. After a day on the wall in the spring sun, it was hot. My wife had never been more beautiful to me than when she glistened like a golden pear in that tub on the stone patio, beside a little bed of bloodred *claveles,* Spanish chrysanthemums whose hue cannot be found anywhere else, that also fed on the waters of that well.

The well was the center around which radiated the spokes of our life there. Its failure would have meant the end of the home. All over the south of Europe, I have come on lovely houses fallen to ruin, and beside each, there is a dry well. For me as a foreigner, accustomed to American plumbing, it was some time before it occurred to me that the people had not just turned off the water when they left the house. The water had left them.

Yet as precious as a well is, it is also dangerous. The neat *aljibe* of our little house had a solid steel cover with a lock on it. At first we thought this was to keep people from stealing the water. It turned out it was meant to keep the well from stealing children.

Our steel bucket told me what it would be like to fall into the well. Not only was there the erotic and scary period of two or three seconds before the bucket hit the water, but it hit with a loud smack. Before we learned better, we dropped the bucket bottom-down and so destroyed it within a week.

No hole in the earth is a thing to be trifled with. Each leads to the subterranean waters. In 1980, a Texaco rig, drilling for oil beneath Lake Peigneur in southern Louisiana, accidentally penetrated the labyrinthine tunnels of a salt mine. Within eight hours, the entire lake had drained into the mine, taking with it eleven barges, a tugboat, more than seventy acres of land, and the drilling rig itself.

Next day, when the tide rose in the nearby Gulf of Mexico, nine of the barges popped back out of the ground as the lake refilled. The tugboat, the rig, and two additional barges remained underground, where doubtless they will someday delight and dismay an archaeologist studying the ancient twentieth century.

This grotesque event is a caricature of the worldwide soil processes by which life is oriented and maintained. There is in fact hardly a single spot on Earth that is not honeycombed with tunnels, some filled with water and others filled with air. The average channel has a diameter smaller than a human hair.

In the top layers of the soil, at least fifty percent of the channels are filled with air, the rest with water. Beneath these rests the "saturation zone," where the groundwater fills all the channels. This zone is also called the "water table," a poetically appropriate piece of scientific nomenclature that calls attention to a simple but often forgotten fact: We are all basically afloat.

The difference between a desert, a fertile field, and a swamp is not the presence or absence of water in the subsoil, but its availability. Free, loose exchange from depth to surface—the channels narrowing gradually from depth to surface—is the best recipe for a fertile soil. When it rains the water diffuses through the deep layers, from which it can be drawn upward again by capillary action as the soil surface dries.

The trouble comes where the water table lies too far beneath the surface. Then, no matter how beautiful the structure of the surface soil, it has not enough sucking power to draw the deep water up. Or perhaps a layer of tough, impermeable stone cuts off the deep water from the surface soil. In either case, the farmer or the homesteader is unhappy.

When human population pressure was less, people simply left these areas alone. Not anymore. For the past two hundred years, the globe's lands-of-little-rain have come under increasing settlement pressure. Deep wells have become increasingly important. And with these pressures has come a class of people who claim to sense directly where the best, most easily available water is to be found underground: dowsers.

Dowsers pay attention to a group of sensations that in most of us have simply atrophied. How does dowsing work? "We live in a universe of waves," says Terry Ross, a dowser from Vermont. "And we ourselves can generate waves that resonate with other waves. At a certain level of resonance, you get a response." He has used most of the dowser's tools—Y-rods, L-rods and hand-held pendulums—but he is convinced that the material is not crucial. "The rod is just a dial on the stick of the mind," says Ross.

Essentially, dowsers claim that they can find you the sweetest water at the shallowest depth on your property. They do this by walking around holding some simple tool until the device indicates where the best, nearest water lies. There actually *is* a pull on the rod as you walk across the land with it. I myself have felt it happen, though I could not tell you whether or not I'd found upwelling water there.

Dowsers are aware that the Earth is a living being, with water circulating under her skin and rippling magnetism girdling her inside and out. Soils themselves are complex magnetical bodies, influenced by the size, the source, and the age of the soil particles. John Ruskin wrote a remarkable essay called "The Work of Iron," in which he praised the colors that oxidized iron produces on the Earth's surface. Invisible but more influential, perhaps, are the webs of magnetism transmitted through the same iron in stones and soils.

Indeed, William Gilbert, the discoverer of magnetism, was deeply impressed by the fact that though it could easily be measured as a field, it had no physically quantifiable matter. If you took a magnet and cut it in half, for example, each half did not possess half the magnetic field. Each possessed the original field.

To Gilbert, magnetism was to the Earth as a soul is to the human being: massless and invisible, yet more essential and indivisible than the physical nature. He knew that the whole Earth was a magnet, owing to its rotation upon an axis, and he connected this rotation to the pattern of day and night, thus making magnetism a co-creator of everything alive.

On Gilbert's Earth, it would be little wonder if man, bearing copious iron in the haem of the blood, had not a magnetical sense capable of orienting itself to magnetic anomalies in and on the Earth. It is a well-known fact that birds, turtles, and other animals have such a sense. A homing pigeon taken a thousand miles from its roost and blinded—what an awful experiment!—will nonetheless find its way back to within a few feet of its native roost.

One scientist whom I interviewed about dowsing scoffed at the very possibility. Without prompting, however, he noted that animals studied in China do seem to be excellent advance predictors of earthquakes.

Nature, theologians say, is unlike man, in that she is not fallen. Animals and plants still experience the world as our first parents might have: not only for its visible and sensual properties but also for its invisible ones. Science tells us we are lords of Creation and that we know everything, but it would seem that our mental world is often more impoverished than that of an ant or a weed. The compass plant orients its whorls of leaves on the North-South axis.

We do not know the simplest things. We have given up a lot in order to be know-it-alls. No one can conceive today how the pyramids were built, how Giotto mixed his colors or how he drew a perfect circle freehand, or how a tribe from the Polynesian Archipelago sailed repeatedly across open ocean and each time reached the Island of Hawaii, a feat rather like hitting the bull's-eye with a BB from a hundred miles off. It isn't that we have lost the technology. We are simply missing the inner resources that make for real craft.

I wrote an article about dowsing for *The New York Times.* The editors were vexed that I could provide no scientific corroboration. Readers sent me copies of the latest debunking reports, showing that there was no way dowsing could be anything but charlatanry.

But I have seen charlatans running the nightly news and major medical research foundations as frequently as I have seen them in dowsers' circles. It is a false distinction. And it makes me very happy to think that there are people out there making the effort to get back into their guts the knowledge that, a millennium ago, perhaps whole populations had.

The groundwater is down there deep beneath a layer of granite. Mr. Ross is looking for a hole in the granite. I hope that he finds it. And I hope this only as a particular case of the hope that embraces other invisible findings as well. I hope to credit and to exercise more the sense that tells me for sure when a loved one is thinking of me, that unheralded sense of well-being that invades children, or the certain sense of disaster that sometimes causes me to avert it.

The motto of modern dowsing is "Indago Felix," or "the fruitful search." It wouldn't be a bad motto for any life.

V

Clay Alive

IN A LANDSCAPE OF CLAYS

INERT MATTER! As if there ever were such a thing. Beauty is the vocation of the world. Under the electron microscope, one can perceive the intricate harmonies of clays.

A Hawaiian imogolite clay looks wilder than the lava fields at Mauna Loa. It is a labyrinth of tubes into which molecules disappear.

A Chinese kaolin is more complex than the karst mountains of which Li Po sang: its particles are long hexagonal bars, piled side by side; millions of parchment sheets, each curled slightly at the edge.

An Iowan kaolin is more gnarled than a bluff along the Mississippi. Its hexagons stack like ancient limestone formations. A particle of an illite from Missouri looks like fireworks, or like the corona of the sun during an eclipse.

A Spanish illite curls back like the tongues of cats. The montmorillonites of Arizona and Wyoming resemble dense growths of watercress. Here is one like a cupboard full of plates, another like a bowl full of needles, a third like the capillaries of the lungs.

The assemblies of the clays are like those hedge mazes and forests in which fairy-tale children become lost, like those places where the old woman is met and where treasures are won. The landscape of the clays is like the wall of the stomach or the tree of the capillaries, or the intricate folds of the womb. It is the honeycomb of matter, whose activity is to receive, contain, enfold, and give birth.

My favorite medieval saying is about Mary the mother of God. "What the whole world could not contain," so it goes, "did Mary contain." There is more real sex in that one sentence than in all the so-called erotic literature ever penned. And it is exactly about the principle of matter, whose activity is fully and willingly to receive.

THE THEORY OF SILT

WE THINK THAT IT IS THE BIG OR THE LITTLE THINGS that make the difference in life: the biggest dam ever, the smallest hole. But the persistent middle term—the stuff that just keeps coming—that is what really changes the landscape.

Clay settles, sand hops, silt slides. Rubbed between two fingers, dry silt feels lubricious. It is the smoothness of a soil. Silt is the middle term between big sand and submicroscopic clay; it mediates between chemistry and structure. It is the soil when it daydreams, pushed this way by the water or that way by the wind, present but not decisive, flighty but not unhinged.

With the soil, as elsewhere, we usually forget the middle term, focusing instead on the extremes. But it is silt that most often travels, and so it is silt that is responsible for the best and worst that soil can do.

If you drop a particle of coarse sand in water, it will fall about four inches in one second. A particle of very fine clay, on the other hand, will take about 860 years to fall the same four inches. Silt will fall the same distance in five minutes.

Silt flew from the trailing edge of retreating glaciers to make the fine black soils of the Midwest. It slides along the streams of the world, leaving deposits that make whole regions fertile. But silt, too, led the other particles north to Canada, when the dust storms of the Great Depression came.

Other conditions being equal, soils rich in silt are the most erodible. Raindrops striking the soil surface and running in the plow furrows dislodge silt particles from their dry resting place, knocking them into the pores through which the soil receives the water. The much smaller clays then pile up around these partial blockages, and soon you have a thin crust on the soil surface that prevents water and air from getting in and germinating seeds from getting out.

It is like what happens when you are rinsing off the dinner dishes and the water runs out through one of those perforated catchers that keep the drain from clogging. The spaghetti sauce's smaller particles pass, until a couple of hunks of onion settle in the catcher; after that, more and more particles are impeded and a cap develops. You curse and get your sleeves wet remedying the situation.

Much of the disastrous erosion of the American Dustbowl began when rain struck unprotected soil, rearranging the silt and clay and starting sheets and gullies of water running across the soil surface. The winter of 1932 was a wet one in Mississippi's Yazoo River basin. Twenty-seven inches of rain fell, and the flooding was disastrous. Sixty-two percent of the runoff came from bare fields, carrying with it thirty-four tons of soil per acre into the stream. In a nearby oak forest, a little more than an eighth of an inch ran off and only seventy-five pounds of soil per acre were lost. Tree roots are not the reason that the soil is held in place, so much as the tree and plant canopy, including the litter layer with which it covers the mineral soil, is a shield that preserves structure. In Cooper Basin, Tennessee, where smelter fumes had destroyed a hardwood forest, leaving bare soil, the maximum runoff increased from thirty cubic feet per second to 1,263 cubic feet per second. Fires also increase runoff, not only by destroying ground cover, but also because chemical changes in the surface give the ground a negative charge, causing it to repel water molecules.

But this is not silt's fault. When it is functioning as it ought, it dreams in the rivers of the world, forming the fertile deposits of their floodplains and their deltas. Lifted into suspension in the faster upland waters, it remains there, until in the lowlands, like a glider coming in for a landing, it bumps to a stop. In the dozen major rivers of the world, it accounts for more than four and half billion tons of floating matter each year. To equal the silt output of the Mississippi River alone, you would have to run almost 25,000 railroad cars full of silt to the Gulf of Mexico. Daily.

Until recently, silt's annual deposition along the banks of the Nile River accounted for the longest-lived prosperous agriculture in the Western world. Then, in 1980, the Aswan High Dam was opened. Like any dam, it arrests silt. Velocity and gradient drop to zero at the head of the dam's still-water zone and the silt settles, falls out of suspension, begins to back up. It is not a good end to silt's dream.

In fact, the invention of dams has changed the word "silt" from a noun to a verb. Now a dam is said to "silt up" or to suffer "siltation." A little dam like the Mono Reservoir Dam, built in 1935 in California, can silt up quickly. The steep bare landscape around it had been denuded by a forest fire. In two years, the silt reached the top lip of the dam. It will take Egypt's Lake Nasser very much longer than that—perhaps centuries—to fill, but someday it will.

In the meantime, the water runs clear and pure from the Aswan High Dam, after spinning through the turbines where it generates ten billion kilowatts of power per annum. The loss of the silt downstream, however, is devastating. A million acres of farmland now must be fertilized with industrially produced amendments. Furthermore, the clean water is no longer saturated with sediments, so it gathers a fresh load of mineral matter downstream, increasing erosion below the dam and requiring the total reconstruction of its infrastructure, including four smaller dams and over five hundred bridges. Because sediments no longer reach the sea, the delta is eroding at a rate unprecedented in four thousand years, and the catch of fish, dependent on the nutrient-rich sediment that fed invertebrates, has declined by more than a fifth.

But silt will have its way. Eventually, it will dream itself even to the summit of the Aswan High Dam, and the great project will become a spectacular waterfall, a natural wonder of a different order.

CREVICE INVASION

AN ORANGE LICHEN CLINGS TO A CRACK near the base of a granite boulder. It holds a little rain, catches dust that has blown in from the Great Basin, and begins to secrete acids, which work into the grain of the rock. Water makes a wedge in the crack, and following it, a birch seedling sends out its root along the edge of the fracture, living on the products of its leaves' decay and the few minerals that the water and the lichen have liberated from the rock.

Water tears rock, and so there begins to be soil. Sir Isaac Newton, in a book on alchemy, *De Natura Acidorum,* suggested that all substances could be reduced to water. Since his time, scientists have found the basis of matter in far smaller units than water, yet it might still be said that all substances are reduced to their own fundamental parts *by* water.

Water gets into things. It soaks them, drenches them, permeates them. No watch or coat is truly waterproof. There is no legal definition of the word "waterproof," because there is no such thing. Nothing resists water indefinitely. Even my Stetson hat, which features a picture of a cowboy using the hat to water his horse, soaked through after four hours in a driving Pacific storm.

The same thing happens to rocks. Some are porous to start with; some contain elements, like calcium carbonate, that dissolve in water, creating channels. Sometimes a mechanical shock starts the process of disintegration: a moving fault or a crashing wave produces a microscopic fracture in the crystalline structure. A thin film of water insinuates itself into the crevice. From that moment, soil becomes possible and, with it, life.

Water behaves strangely when it gets into a crack. In that environment, it is called interstitial water, vicinal water, orthowater, or ordered water. The key thing is that it does *not* freeze.

The thin film of water in the crack starts to freeze, but it can't. When the temperature drops, the water molecules try to migrate, reordering themselves into ice crystals, but they are already in an ordered relationship with respect to the crystals of the rock walls. A tug of war ensues. The tension is enough to pull rock grains apart. Equivalent force, on a larger scale, would be a wind so strong that it ripped the facades off skyscrapers or a pull sufficient to part a bridge cable.

It takes 210,000 pounds of pressure to break the surface tension of a one-inch column of *pure* water. So as invisible as the process is to our eyes, it is nevertheless explosively powerful. The results can be seen deep in the profiles of older soils, where the digger comes upon huge boulders of granite or gneiss that crumble at a touch. They perfectly resemble the parent rock, but water has eaten out their structure from the inside.

In nature, however, water is seldom pure. When salt-impregnated water enters a crack, the salt may gather more water around its strongly attractive ions, a process called "hydration," forcing the crack apart. If, on the other hand, the water evaporates, the salts may crystallize into a solid state again, gradually splitting the rock as the crystals grow. Furthermore, since salts expand more when heated than do most rocks, hot salt will open a breach further.

What does it mean, then, to be as solid as a rock? Better to consider the fragility of rock, and its transformation into soil. One third of the sedimentary rock in the world is derived from clay—all of which is derived from weathering of other rocks. At first, the stone breaks down into smaller and smaller parts, but at a certain point the process of destruction concludes and the fragments begin to build a new structure: one of the earth's indispensable clays. If you spread all this clay evenly, it would make a layer one mile thick over the whole surface of the Earth.

Even human monuments show the delicacy of stone. The 4,500-year-old step pyramids of Egypt are among the oldest human structures known. Each terrace of the pyramids is covered with a thick talus of cracked-off, crumbled stone, much like the talus found at the base of mountains. Each building stone has turned from a rectangular piece to a rounded boulder. The Great Pyramids, too, are beginning to turn to boulder piles. Up until a thousand years ago, when their polished facing stones were carried off to Cairo to use in building mosques, the core of the pyramids was protected. Since that event, the more vulnerable limestones among the blocks have been losing matter at a rate of fifty cubic centimeters per year.

The castles and churches of Europe promise to be even more ephemeral. A nine-hundred-year-old castle in Austria, made of sandstone blocks, is a ruin that sprouts grass from the clay soil that has formed atop what once were its walls. A 770-year-old church of the same sandstone is deeply weathered but still intact. A five-hundred-year-old church is just starting to show wear. And a one-century-old church is as fresh as when it was built.

Limestone and even marble tombstones weather so fast that the memorial to grandfather may scarcely outlive those who knew him, disappearing like the writing on a magic writing tablet after you have lifted the top page. When polished marble weathers, it first gets reddish iron oxide streaks, then the stone roughens as the polished surface grains are split from the underlying stone; finally, the grains form a sugary coating that can be be scraped off by hand.

The landscape of the whole Earth is itself little more than a monument to the different weathering rates of its constituent minerals. Peaks and valleys, beds and swales all result from the variable weathering of stone into soil.

Generally speaking, sandstone is quickest to go, then marble and schist, then granite and gneiss, so the resistant granites occupy the mountain heights, the middling-resistant marbles the uplands and slopes, and the quick-melting sandstone the valleys. In an Edinburgh cemetery, the marble tombstones lost three and a half inches of their surface in a millennium; a granite stone lost only one tenth that amount.

The sum of the processes of destruction can work so fast that they become perceptible within the scale of a human lifetime. In 1881, Captain Henry H. Gorringe brought an elaborately carved Egyptian memorial obelisk from Alexandria to New York's Central Park, where it was set up with fanfare and dubbed Cleopatra's Needle. Within a year, the beautiful hieroglyphs of the west and north faces were already fading; after three years, piles of flakes could be collected at the needle's base. Eighty years later, the west face had lost one and a half inches of its surface, which effectively erased its entire inscription.

The monument was made of a hard but coarse-grained syenite granite, quarried in Aswan, where only one to three inches of rain fell per year. There, the stone weathered at a rate of only one-half inch in two thousand years. In New York, where forty-three inches of rain fall each year, you might expect faster weathering. But salt and temperature are the more telling agents. The obelisk had lain for several hundred years on its side in the Nile silt, where Persian invaders had toppled it. During that time, the muddy faces had imbibed the natural salts of the Nile.

When Captain Gorringe set up the salt-impregnated stone in rainy New York, the increased moisture rose through the base, forcing the hidden salts to the surface, where together with the immense pressures of freezing and thawing, they made short work of the inscribed names and the mighty deeds of Queen Hatshepsut, Thutmose III, and Ramses II.

CLAY AND LIFE

AT CHRISTMAS, I was out on the prairie again. Third time in a year. It seems like I can't stay away. This time, I came up out of Council Grove, Kansas, at seven A.M., just around sunrise. For about a minute and a half, I saw the sun and the full moon balanced evenly at the opposite ends of the sky. And here was I, riding along the bald and slightly arched surface of the earth, halfway between the two.

What are we doing on this planet, and how did we get here? It took only a glance to tell that there would not be anything like us found on the yellow sun or the fast-paling moon. The Earth has one thing that neither sun nor moon has ever had.

And that one thing is clay.

I stopped by the Spring Hill Ranch, where there's a break in the fence. I like to walk the erosion gullies on the virgin prairie. There are fossils in the lower strata, as thick as nuts in an almond bar. But that morning, I found clay. I dug it out with a stone and formed it into small flutes and bowls. It was almost greasy to the touch, it had a sheen about it, and you could shape it into anything.

What if clay is alive? Biologist Hayman Hartman thinks it is. If he is right, then perhaps our ultimate ancestor really was *adam,* the Hebrew word for "red clay."

"Our difficulty," says Hartman, "is that we think of organic molecules as somehow being the necessary and sufficient condition for life." But what if life were a more pervasive phenomenon than that? What if life were already implied in the magnificent, supple matrices of the clays?

The word "organic" comes from the Greek *organon,* meaning "tool" or "instrument," the latter itself derived from words that mean "to do work" or "to perform a sacrifice." By these definitions, clay is at least as lively as the so-called organic chemicals. It may quite literally have been the matrix (the old word for "womb") that spawned all the creatures now inhabiting the earth.

In 1953, chemist Stanley L. Miller shot a tankful of inorganic chemicals, designed to simulate the ocean, full of simulated lightning bolts. He succeeded in synthesizing simple organic compounds, thus giving rise to the "organic soup" theory. The idea is that life on Earth began when inorganic chemicals were bombarded with electromagnetic radiation. Presto! Life ensues.

Cosmically speaking, however, it is all too easy to make organic compounds. Hydrogen, carbon, nitrogen, and oxygen—the chief components of organic compounds—are the commonest elements in the universe. Microwave investigations of the deepest galactic clouds reveal the presence of the ingredients of organic chemistry. Compared to them, the Earth is poor in these crucial compounds. So why should it be that the Earth, alone of all the bodies with which we are acquainted, should harbor what we call organic life?

The mother function is apparently far more active than Miller (and many of our parents) expected. Indeed, it is the crucial factor in the equation. To "spill your seed on the ground" is to spend your energy where it receives no answer. The flaw in Miller's experiment was to think that the ocean was anything like a retort. Even had small organics been created in it, they would almost certainly have been churned to pieces before longer and more lifelike chains could build. There was not enough motherly repose in the open sea, nothing to enfold, contain, and order.

This is exactly the function of a clay. Formed by water, it is the seat of a wild, capacious order. When the physicist Erwin Schrodinger speculated on the fundamental building blocks of life, he concluded that the basic component would have to be an "aperiodic crystal," that is, an ordered and repeating structure that nonetheless left room for a whole variety of actualizations. Only such a machine, he reasoned, would be sufficiently supple and dynamic to engage in the constantly shifting behaviors of metabolism, growth, and reproduction. DNA—discovered decades after Schrodinger's assertion—admirably fulfilled these conditions. But so does clay.

The clays, unlike their parent rocks, have no inaccessible interior, but instead a very large reactive surface. They unlock the potential waiting in raw matter. Clays stack, wrap, pile, and exfoliate, like leaves or sheaves of paper. In fact, experiments have shown that a single gram of a clay powder can have a total surface area larger than a football field. It is as though you set out to write a book and had to decide between writing on a ream of paper or on a raw tree. Which to choose?

But clay is more than an empty book. It is already encoded with a vast array of possible meanings. Each layer of a clay is a matrix of molecules that is well ordered but marked with a succession of dimples, holes, and ragged edges. In other words, each layer is a template with tendencies.

Isaiah, describing how God would cause righteousness and praise to arise, compared the act to a garden, which "causes what is sown in it to spring up." The ground itself is as active as the seed. The seeds of organic life, attracted to the patterns of a clay matrix, might well have found there the structure that makes all of us possible, and the means to maintain and reproduce it.

Yet do clays actually behave like this? Direct evidence is lacking, since even to measure the order in clay crystals requires a scale of microscopy that is only now beginning to become available. X-ray diffraction technique and the scanning electron microscope, technologies that permitted our first glimpse in the macro world of the clays, are too blunt as tools to assess and compare the levels of patterning in clays. The new atomic-forces microscope shows promise for unraveling the complexity of this micro level of clay order.

Was clay, then, the necessary partner for the birth of the organic realm? Hartman says, "There are only two things in the universe that require liquid water for their existence: organic life and clay."

The oldest rocks that we can find on earth are a mere 3.8 billion years old, nearly a billion years younger than the earth itself. It is impossible therefore to find a "fossil" record of the earliest clays. But there is a type of meteorite, called a carbonaceous chondrite, whose average age is about 4.5 billion years, comparable to the Earth's. Certain of these meteorites contain both liquid water and iron-rich clays. The same ones also reveal the presence of amino acids and other complex organic compounds. On the other hand, those that contain no liquid water reveal no clays, and those without clays have no organic compounds. Since each of the meteorites evolved in isolation in deep space, there is evidently a connection between the water, the clays, and the growth of long-chain organic molecules.

In its first billion years, Earth was a rather different place than it is now. Oxygen was rare, and iron was common and very reactive. Nothing grew on the surface of the planet, so the land flowed downhill into the broad shallow shores of the seas. Most clays formed there or near the seafloor vents, where fresh iron was pumped upward from within the crust.

Still today, in deep-sea trenches largely deprived of oxygen, clay species rich in iron and potassium form and evolve within the pore spaces of seabottom minerals, relatively protected from the surrounding events. These clays closely resemble what clays must have been on the early Earth. And they behave as though they were alive. They go through a process of transforming growth. A young one resembles a little worm, and blooms into a pattern of open curves. Another begins as an isolate thread, and grows into nodules that some scientists have compared to peppercorns but that also have the comb structure of a beehive or a sponge.

It might be argued that these patterns are nothing more than crystal growth. But the real question, as Schrodinger knew, is how life itself is related to crystal growth. Richly patterned clays might have served as templates for biosynthesis, that is, for the beginning of organic life. It is a statement so simple and obvious that it runs the risk of being ignored, just as the obvious case for plate tectonics was ignored for hundreds of years.

Consider the iron-rich mineral called kakoxen. Tubular in form, hexagonal and hollow in cross-section, this remarkable crystal species looks like nothing so much as the rose window of a Gothic cathedral.

Exactly the same impression is given by computer simulations looking down the bore of a DNA molecule. Seeing both of these together, it is easy to believe that the protected interior of such a crystal might have been the site where organic polymers of the sort that would form nucleic and amino acids were born.

Hartman has an idea how the transition from these probiotic clays to organic life might have occurred. In the presence of ultraviolet light, iron can sometimes capture both carbon dioxide and nitrogen from the air, resulting in the production of citric acid, an organic compound. Amino acids can gradually be built out of citric acid. So seaborne clays on the ancient Earth, deriving their energy by feeding on CO_2, nitrogen, and light, might produce the building blocks of organic life.

A basic question for Hartman and his colleagues is this: Do (or did) clays reproduce? In other words, does the particular order expressed in one particular clay create further clays with the same order? Second, is this order dynamic? Does it lead to social interactions among clays?

It's an exciting question, but possibly a blind alley. After all, the mythology of heaven is full of living creatures—the dwarfs, the fairies, the angels—who do not reproduce as we know it. Perhaps, instead, we could see life as a strategy for fulfilling the wishes of a given matter—whether molecular, cellular, or psychological—endowed by the creator with a constellation of possible forces. Whatever the matter may be, life is neither chaos nor prison, but process and freedom.

The crystallographer A. L. Mackay notes that any lifelike system requires "a stream of energy [that] passes through the system and its environment." Life begins in this interaction, where the energy is bent and diverted into little chaotic vortices, unexpected patterns, dynamic containers of information. A clay crystal, he says, fulfills just these requirements. He compares it to an abacus, well ordered but capable of many meaningful permutations. A certain minimum energy is required to change it from one state to another. It is therefore a code.

The clay code, however, is more complex than either the genetic code or human language. Only now are we beginning to catch glimpses of its order, and one cannot help thinking that pursuing it will be as fruitful and as endless as the cabbalists' search for that perfect expression of the Hebrew *aleph,* by which God created the universe.

It is said in Genesis 6 that once upon a time the sons of God came down to earth and begot children on the daughters of men. If we admit that clay is alive, we must say that it is both more ancient, more widespread in the universe, more durable, and more powerful than we are. Yet it is also less supple and less able to make abrupt transformations. Perhaps *this* Genesis story can symbolize the rise of life as we experience it, from the joining of organic and inorganic realms. Wouldn't it be strange if, in the history of the living, clay performed the function of angels?

KAOLIN

IN THE EARLY PART OF THE TWENTIETH CENTURY, archaeologists discovered Minoan Mycenae, and Crete, the palace of Knossos, the remains of the Minotaur's labyrinth, and several thousand hunks of dried clay, all inscribed with a beautiful but unreadable script. It was the heyday of the discovery that these cities, described in myths, had really existed on the Earth. Troy, Mycenae, Pylos had not been products of the fevered Homeric imagination, but of actual histories and events. Imaginations began to awaken.

The archaeologists who sought to decipher this odd script, most of which is called Linear B, were ready to find anything. Looking at one big disk of inscribed clay, F. G. Gordon translated it as "The Lord walking on wings the breathless path, the star-smiter, the foaming gulf of waters." Another archaeologist translated it as "Arise, savior! Listen, Goddess, Rhea!" A third made it "Supreme—deity of the powerful throne's star. Supreme—tenderness of the consolatory words."

In fact, however, it eventually turned out that the vast trove of written clay said much more prosaic things, like "Thus the priestess and the key-bearers and the Followers and Westreus hold leases: so much wheat 21.6 units." It was all records and inventories, storage documents of the sort one might find in the office at any city warehouse or country grain elevator.

But without the ability to contain, as well as produce, no culture can be created. To the inhabitants of the first cities, a warehouse *was* a miracle. The first discovery of settled life was that it was possible to save. To preserve and to transport what had been preserved, it was necessary to have containers. A cave had to be found; a house could be built. A niche stayed where it was, but a pot could travel.

The first free-standing containers, whether houses, fences, or baskets, were doubtless made of wood, straw, and leaves. The materials were easy to work and to shape with the simplest tools. Neither, however, was as impermeable as one might wish. The house lets in the wind and the cold; the basketry let in the damp and the mice.

No one knows when a change came, but it appears to have occurred at about the same time that alphabets were invented. Perhaps a woman had taken to lining her baskets with clay, in order to improve their water-holding capacity. Maybe she accidentally dropped the basket in the fire and removed it later to find that the wood had burned away, leaving a hard and apparently permanent container of clay, the first pot.

I prefer to think that she got the idea. Certainly, the children would have played with clay, delighting in its plasticity. Clay is found everywhere on the surface of the Earth, and is particularly exposed along river banks, where the water cuts away the topsoil layer in its steep banks. Mothers and children, at least, would have spent a great deal of time by the river. Maybe a child, in imitation of basketwork, made the first pot out of clay. At the end of the day, her mother, coming to retrieve her from the bank where she'd been playing, took from her daughter's hand this odd basket-that-wasn't. She thought, Why not?

There were probably many such scenarios played out in separate small encampments wherever humans were beginning to settle, but each depended upon clay. And not one culture developed square pots. Pots and plates were broad and bellied objects, rounded like pregnant women. They were the containers of the precious principle by which a people might continue into the next generation. They were the definition of prosperity, which means, "to hope towards."

Clay is plastic because it is made of infinitesimal plates that slide one across the other, held loosely in place by intervening layers of chemically combined water. It is very hard to pull the plates apart, but comparatively easy to slide them one across the other—as anyone who has tried to walk in the sticky gumbo-till clays that sometimes cover the surface of the Earth will tell you. It can be almost impossible to lift your foot out of the sucking mass of clay, while when you try to slide it forward, you may find it shoots ahead, leaving you sitting on your rump.

Why are some clays red, some white, and some brown, some even black? Every clay is simply an evolved expression of the Earth's crust. It contains in corresponding proportions the important elements that go to make up crustal rock: silica, alumina, iron oxides, and in smaller quantities, oxides of manganese, calcium, potassium, and other elements. Long weathering has reduced these elements as far as water is able, making the particles very small and dissolving whatever compounds could be dissolved. The composition of clays determines their color.

All the rich colors come from the earth. They have to do with the temperament of iron. Pure sand is pure silica: it is white. Alumina is also whitish. The purity of porcelain dishes is owed to the fact that few impurities exist in them. They are entirely silica and alumina, the elements that together make up more than three fourths of the bulk of the Earth's crust. But even on porcelain, this white is merely the canvas against which the theater of the warm and the cold colors comes to life.

Iron acts in clay as it does in the blood. When the blood is oxidized, its iron-rich hemoglobin is red; when the blood is poor in oxygen, the hemoglobin turns blue. Every potter knows that the same distinction exists among the clays. Any clay that contains iron—that is, almost all but the pure kaolins—will fire into a warm, red, yellow, or red-brown color, so long as there is plenty of oxygen in the atmosphere of the kiln. If you deprive the kiln of oxygen, the clay will fire to a cool color, a blue-green or a black or a deep purple. On the surface of the Earth the same holds true: clays exposed at the surface or in a porous soil run the whole gamut of reds, pinks, yellows, browns, ochres, but a waterlogged clay along a stream or a bog is a cold greenish-black.

The fresco painters of the Italian Renaissance found themselves in a peculiar position with respect to color. They had available to them a large number of vegetable- and mineral-derived pigments, but the technique of *fresco* (that is, working on wet plaster) limited them largely to the earth's palette, because the alkali in the plaster tended to decompose and disperse the vegetable-based dyes. The very rich colors of Massaccio's frescoes are almost all derived directly from the soil. The reds, browns, and yellows are from ochre. The green is from a reduced clay called *terre verte*. The umber came straight from the earth of Sienna. The whole Christian drama is expressed in the colors of the earth.

THE PATH OF A CLAY CRYSTAL

I USED TO CLIMB IN THE SIERRAS. I was not such a good climber, but I was good enough to get into trouble. I was an apprentice teacher of rock climbing in a camp that was based at ten thousand feet near Mount Langley. Our students that year were high-school kids from East L.A. They were tough but good kids. And fooled kids. The recruiter who'd snared the guy on my Mount Whitney rope had told him he was going to the mountains to go water-skiing! (He'd confessed this to me just before I pushed him over the edge of a 150-foot drop for his first rappel.) He was big and strong, formidable on his own ground, but a baby at the end of a rope.

We'd got on the wrong route, not the easy broad-ledged pitches around a big outcrop we called the Bus, but up the scary vertical and tilted pitches of Shaky Leg Crack. There was about two thousand feet of clean granite exposure stretching downward in a gentle concave arc beneath our feet. That was when my ropemate said he couldn't go any farther.

What was I going to do? I didn't know if I could go any farther either, but I knew that he had to. So I took his rucksack off his back to lighten his load, tying it off so that it hung by a hank of rope from my waist. Now, I was quite certain that I couldn't make it, even if we found the right route again.

Stepping up to lead the next pitch, I came into a wide shallow crack that leaned a little outward from the face and to the right. Again and again, I tried to get wedged into it, only to fall back. I was terrified. It was clear that the only way to do this move was to turn face out and shimmy up it. The dangling pack was in my way; my ropemate was saying he couldn't hold me; and I was, I felt, unquestionably about to fall.

Perhaps thirty seconds later, without knowing exactly how, I had made the move and found myself jammed in the crack—by the shoulders, upper arms, hands, knees, and feet—looking straight between my legs at a little alpine lake way below me, and out across the broken skirts of the Sierras, across the Owens Valley and to the high, stark White Mountains on the other side. All the fear was gone from me. In fact, I felt as though what became of me was a matter of no great consequence, and that regardless, I was completely happy. The question as to whether or not I would fall had become irrelevant. I felt I saw the world—in that moment—as it actually is.

Aquinas says that God is in the world not as the essence of all things but as their cause. That, I think, is what I saw that made it possible for me to relax more deeply than I ever had before or have since. The divine was not some Thing in which to "believe," but living and active, not far off and deigning to descend, but the common principle of existence. It filled everything, yet could be diminished by the death of nothing.

All of that time is living to me: the colors that were not stronger but more fully modulated than I have seen since, the interaction of the wind and the sweat on my arms, the shifting weight of the pack swinging like the pendulum on a grandfather clock beneath me.

And with that, I feel particularly the press of the granite around me, and I sense my hands reaching up, looking for little ledges above me, brushing off a bit of the rock's exfoliating skin, smelling it as a dust that for me is still more erotic than the perfume of my wife's sweat.

The divine is the commonest thing in the universe. At that moment, the granite dust and I parted company. My companion and I headed up, straining and puffing, until we reached the suddenly easier pitches of the Grand Staircase. The dust of the rock sucked and twirled in the eddies of the air and went to join the skin of the Earth.

A quarter ounce of plagioclase feldspar clay falls through the afternoon. Half an hour later, my partner and I wait out a brief thundershower under a broad roof in the Grand Staircase. The feldspar is still in the air, lifted and thrown down by the violent air currents, until a raindrop passing at around twenty-five miles per hour, slams into it and carries it onto the fan of talus blocks at the base of the slope.

It slips along among the boulders streaked yellow with oxidized iron. Every block has a weathered rind, more or less thick according to how long it has lain exposed and how much water has struck it. The feldspar slides down with the water through the talus, into a runoff stream and thence into Mirror Lake.

Swept through the outlet, and out into a rapidly descending creek, it bounces down to the Owens Valley, where it rubs off in the bend of a stream meander. (You can see how a creek meander works by looking at the bottom. Where only the heavier gravels can be seen, the current is swiftest. Where sand is on the bottom, the stream is slower, allowing the comparatively small particles to settle.) When the stream dries in summer, the speck of feldspar is blown farther downhill until it comes to rest in the soils created over thousands of years by previous generations of these forces.

The soil is rich in acids, largely generated in humus or through the reaction of water with the high percentage of carbon dioxide in the soil atmosphere. These react with the feldspar, gradually releasing excess silicon and potassium, substituting hydrogen for some of them, and converting the crystal from a tight nest of polyhedra to the alternating sheets of an illite clay.

The special property of the clay is that along its exposed edges and its outer surfaces, it is not electrically neutral. It therefore attracts to itself oppositely charged particles, many of which are necessary for organic reactions, including potassium, calcium, and nitrates. From a locked closet of the mineral world, it has become an open shelf, where the roots of plants may shop for what they need.

VI

In the Dark

ON GOPHER HUMPS

LIFE SATURATES THE POROUS SURFACE OF THE EARTH. Indeed, life's fundamental property seems to be that it spreads. Disseminating its information in the form of discrete creatures, it entails the power both to adapt to new environments and to reproduce both rapidly and fully to occupy suitable ones. Hans Jenny once added up all the estimates of microbial, invertebrate, and vertebrate life underground. The calculation showed that under each acre was a biomass equivalent to at least ten draft horses. There was more living matter beneath the surface than upon it.

A great deal of this pulsing life is invisible to the human eyes. The several billions of bacteria per average square yard escape our notice, though they are responsible for the maintenance and healing of the soil.

Even the comparatively outsized invertebrates may go unheeded, until they prick us somehow. People working in peat soils around the world often complain of a persistent itch or prickle where their bodies come in contact with the soil. Chemical explanations were common—it was thought that some sort of alkali irritant was present—until someone bothered to put the offending soils under the microscope. It turned out that they were full of tiny spears excreted by freshwater sponges whose presence had until that time not been suspected.

Just as annoying but a great deal more visible than such mini-varmints are the giants of the soil world. Biologists picturesquely refer to them as the "fossorial species," meaning the hole diggers. They include moles, ground squirrels, rabbits, marmots, badgers, mice, shrews, armadillos, certain tropical rats, marsupial moles, prairie dogs, and gophers.

The majority of these are rodents, whose adaptation for digging is extraordinary. Their compact, short-legged bodies are like augers with paws, centrally powered and headed by a set of jaws that must chew to live. As anyone who has kept a hamster knows, the large front teeth never stop growing. It is not that rodents necessarily like to gnaw, but if they fail to gnaw, their teeth grow round until the jaws are sutured shut. Therefore, they gnaw. Furthermore, their teeth have the interesting property of growing enamel only on the front side, so that as they gnaw, they maintain an excellent chisel edge, perfectly adapted for more and more efficient . . . gnawing.

A rabbit or a mouse is quite content to gnaw in place. Once it has dug its burrow, it is satisfied. Not so the gopher. Back in the mid-1960s, a correspondent sent the *Journal of Mammalogy* a three-page report that should have won him the Nobel Prize for effort. Not only did he succeed in starting and ending a six-month, open-site experiment with only one gopher, he also managed to follow the gopher around.

On capture the beast weighed five ounces, about the weight of an average peanut-butter sandwich. On a nice fall day, the correspondent set the diminutive gopher down in his vegetable garden near Logan, Utah, gave it a pat on the haunches, and sat back to observe.

He might as well have released a chainsaw. Within fifteen minutes, it had drilled through eighteen inches of soil. An hour later, it pushed up a mound three feet away. After a week, it appeared on the other side of the road, about a hundred feet from the start. And so it went. With the mammalogist in hot pursuit, the gopher covered 390 feet underground before the soil surface froze solid in December.

Undaunted, the gopher waited in his warm underground burrow until the thaw, while the scientist presumably sipped hot chocolate. In February, fresh mounds began to appear two hundred feet from where the animal had last left off. The experiment ended on the twenty-eighth of February, when the gopher was found dead on the surface.

It might be speculated that the varmint spent all his energy looking everywhere for another gopher with whom to create still more gophers, finally abandoning himself to despair. Nevertheless, the digging statistics are impressive. Almost five hundred feet of near-surface tunnels alone, in a free digging time of a little less than three months. In his most active month, he had pushed up seventy mounds of soil, amounting to about five hundred pounds.

Doing some quick calculations to approximate per-acre gopher density, the mammalogist came up with a figure of thirty. Multiplying that number of active gophers by his prize gopher's statistics, he estimated that a good field full of gophers would move about thirty-eight tons per acre per year. In short, under the influence of a good population of motivated gophers, it would theoretically be possible to move all of Utah into Colorado within a century.

An exaggeration? There is a whole kind of landscape in the western United States (also in South Africa, Argentina, and East Africa) called Mima Mound topography. These mounds are up to six feet high and from seventy-five to 150 feet in diameter, sometimes extending over areas of several hundred acres. Once, it was thought that geological agents had caused this distinctive terrain.

Wrong. It was gophers. Like their cousins, the prairie dogs, certain pocket gophers tend to pile all the soil that they move into these large mounds. Students whose job is to examine this "soil translocation" suggest that the animals only stop moving soil when there is no more soil to move, that is, when they hit bedrock. The size and topography of the mounds themselves suggests that a gopher will keep digging until he has moved all nonmound soils into his regular heaps.

The extraordinary beauty, power, and determination of the gopher cannot be denied. But, as every gardener is then prone to ask, how do I destroy them? By "them," he means not only gophers but also the moles, who, from the horticultural point of view, are simply gophers in Halloween costumes. Both are distinguished by the spongy-to-the-touch tunnel lines that they leave crosshatching the lawn and emerging every twenty feet or so in a nice brown pile of earth.

From one point of view, it is a testimony to the gardener's skill at making fine pasturage that the gophers have favored his lawn with their presence. On the other hand, it is inconvenient and ultimately destructive to the lawn.

The soil defends its inhabitants. I remember hours spent filling gopher holes with water in the vain hope of drowning the animals. One could indeed make the hole overflow, but soon the water would percolate away. We succeeded only in fractionally elevating the water level in the creek. Then, there were the little guillotine traps that promised to impale the beasts in a nest of sharp metal rods the moment they emerged from the burrow. Once we caught a gopher by this means, though mainly we acquired my father's anxiety at having these deadly instruments exposed on the ground, where the curious dog or child might receive a memorable surprise.

Catch 'Em Alive traps never caught 'em at all. Perhaps poisons had some effect, but we never knew. The gophers would come and go punctually, like the rain. In other words, we might have a plague of them one year, and hardly any the next. All the control measures seem to have been executed largely for our own amusement.

Indeed, as children we were always secretly on the gophers' side, and it was with real delight that on occasion we actually saw one poke up his head above the soil, while the adults were setting traps for them in another quarter of the garden. When we filled the hole with water, I often imagined that I was surfing along the front of the advancing water, exploring the underground realm that belonged to gophers, moles, and ants.

The only real specific against gophers is nature. To control the gopher, prosper his enemies. But it is just his enemies that have so much trouble surviving the suburbs. The whole tribe of snakes is marvelously adapted for gopher consumption, and indeed in the wild, they are often the residents of former gopher homes. The favorite constrictor in our part of the world was known as the gopher snake. But while we could easily find the snake in the untouched regions of the coastal hills, there was never a single one in our garden.

Perhaps the best thing to do if you have gophers is to be grateful that you don't have prairie dogs. On the other hand, plagues are often scaled so that they are suitable for us to bear them. A friend lives on a 24,000-acre ranch in South Dakota. He has fields of prairie dogs. Each prairie-dog town is about the size of a small suburban subdevelopment: acre after acre of mounds, each topped with a vigilant whistling demon.

My friend has tried a number of forms of prairie-dog control. One was to place salt licks in their cities, so that buffalo would come and trample them. But the bison don't like the terrain and the prairie dogs don't mind the trampling. Another idea was to try to surround the mounds with quick-growing grasses, to give the predators a way to creep close to the mounds. (Prairie dogs are always sure to denude the surrounding area, probably for precisely that reason.) But prairie dogs are not rodents for nothing.

In the end, my friend's best line of defense has been to talk to all of his neighbors about the importance of ceasing to shoot and trap coyotes and hawks. The rancher's traditional hatred of the coyote has only served his subtler but more powerful enemies. A coyote may take the occasional young animal, but the prairie dog, because of the way it clears large tracts of land, is the occasion for erosion that destroys the life-giving basis of the ranch: the grass.

OF WORMS

Some enthusiastic entomologist will, perhaps, by and by discover that insects and worms are as essential as the larger organisms to the working of the great terraqueous machine.

—George Perkins March,
Man and Nature (1864)

THE TROUBLE WITH WORMS is that they seem to have been extruded. Both the way they look and the way they move are not calculated to endear them to animals that go upon legs. Indeed, so much do they appear to have been squeezed out of somewhere, that in the Middle Ages, it was felt that they were spontaneously generated in the earth.

Perhaps this is because there are so many of them. It has been two centuries since Linnaeus gave up trying to create a whole phylum for the worms, to be called Vermes, because there were too many and they were too various. One suspects that it was also too distasteful. After all, the word "vermin," though most often applied to rats, comes from the Latin for "worm."

What makes a worm? It is slinky, slithery, slimy, soft, blind, and voracious. Most, it would seem, are parasites. There are the pinworms that live in the guts of animals, devouring feces and attaching themselves to the walls of the gut. There are the roundworms that pierce plant roots with their spearlike mouth parts, suck the roots' juice, and infect them, producing stunted plants or misshapen roots. There are tapeworms that grow to a length of more than a hundred feet inside the host; flukes that infect the blood; and leeches that suck it. There are roundworms that burrow beneath the skin and encyst. The slender horsehair worm is supposed to be born when a horse's hair drops in water. And the night-feeding slug leaves its slimy trails on everything.

Then, there are all the worms-for-a-day, the larval forms of insects that are responsible for so many depredations in the soil. There are the wireworms that eat any plant tissue they find beneath the surface. The cutworms, larvae of a lovely butterfly, are deadly to all cole crops, creeping up to the soil surface and munching the roots off at the base. The chafer grubs, the carabid larvae, and all white grubs of lawns are the worm phase of beetles, adapted to infesting pasture and lawn. Fly larvae, too, like the leatherjacket, prey upon plants in the soil.

The horror at these beasts is perhaps fundamentally sourced not simply in their habits but in the fact that they colonize the dead. Disinterments are richest in the larvae of beetles and flies, yet even unburied corpses may erupt with teeming nests of worms.

The contest of man against the worm is most clearly expressed in the legends of dragons, greatest worm of all. Though they are usually reptilian and avian in their external characteristics, their fundamental characteristic is to undulate through the air and over the ground. Furthermore, they have the parasitic voracity that belongs not to reptiles but to the Vermes. (In earlier times, indeed, snakes were regarded as the kin of worms.) Most dragon fighters in Western mythology get the call because a dragon has been demanding a virgin per year, depleting the number of young women, until finally only the princess is left. The worm destroys the principle of increase and motherhood. The hero reinstates it.

Yet the worm is undeniably necessary. It hoards and guards treasures. He is a sort of banker of the fairy-tale world, though he makes it difficult to effect withdrawals. He is in this respect the representative of the dark side of metamorphosis.

The worm takes from the stock of the world's beauty and hides it in his blind life in the earth. He stores it in a body that grows fatter and fatter. He is the embodiment of gross materiality.

In the end, however, he falls into a permanent sleep inside his self-made house. From his sleep, the grub never awakes. Instead, some process transforms the bloated white grub into a creature with colored wings or an iridescent carapace.

Which is grosser? The entombment of a grub in a cocoon, or the swelling of a child in the mother's womb? It would be hard to say objectively. But these ultimate descents into materiality seem to be the preferred locus for miracles.

PERCEPTION IN EARTHWORMS

THE EARTHWORM IS A SPECIAL SORT OF WORM. Almost alone among his brethren, he does not inspire horror. In fact, the earthworm is almost alone among all the invertebrates in the tenderness he inspires. Knowing that a worm in the sun is as good as dead—since his skin has no defense against desiccation—children often place him gently in the shadow of a log or cover him with a light handful of soil. And gardeners, above all, venerate the worm.

The earthworm has also inspired an unusual quantity of study. Charles Darwin, who spent the better part of his time conceiving *The Origin of Species,* devoted a whole volume to the worm. In his *The Formation of Vegetable Mould through the Action of Worms* (1896), a classic of soil ecology and a charming book, Darwin not only summarized all the knowledge then current on earthworms but also reported on his own experiments with exposing worms to light and heat. It was his delight to have friends send him interestingly shaped worm castings from around the world. In his concluding chapter, he begins, "Worms have played a more important part in the history of the world than most persons would at first suppose."

Indeed. Though Darwin's own notions regarding the primary importance of worms in the creation of humus leave out the very important influence of the microbial population, it nevertheless points in the right direction. A healthy population of earthworms is the sign and partly the cause of a healthy soil.

When worms are happy, there are lots of them. In a Danish forest soil, researchers have found a density of one million to one and a half million worms per acre, more than two tons of worms. A rich grassland may bring up more than five hundred worms out of a square-meter hole.

This is not so remarkable when you recognize that eight relatively healthy worms will produce fifteen hundred offspring in half a year's time.

There can even be too many. George Perkins Marsh, writing in the mid-nineteenth century, reported a conversation with an elderly New England pioneer who in her girlhood had witnessed the rise of worms in that part of the world. The common earthworm is not native to America, having been brought over by colonists. When it first appeared, it was not numerous, she reported. But as fields were cleared, its numbers increased to such a degree the water of springs and wells was polluted by the number of dead worms in it. Only the corresponding introduction and increase of robins and other vermivores corrected the imbalance.

Nevertheless, the presence of earthworms is by and large a very good thing for the soil. Unlike a given fertilizer, for example, it acts simultaneously on several different soil parameters.

Worms are basically blind. They see by eating. Darwin tried in vain to scare his worms with a light, but he found that they would only withdraw into the burrows if the light were a hot one. Heat/cold and wet/dry are the polarities that matter to worms. Both are meant to keep them at the optimum lubriciousness for ingesting and passing soil through their bodies.

A worm is a long intestine. Soil, rich in dead organic matter, leaves, and especially manure, goes in one end and comes out the other, concentrated, enriched, and well mixed, in the form of "castings." A well-manured soil is almost always rich in worms. Up to ten tons of worm castings per acre per year enrich a soil under favorable conditions. The worm also senses and creates the topsoil in a very basic way: by going where the organic matter is, mixing it, and excreting it behind or above him. Worms also bore down to the water table, but not into it. At the dry surface, too, they stop. More than any other creature, the worm defines topsoil.

Some leave their castings on the surface, others in the body of the soil, but in any case, the leavings have several virtues. For one thing, they concentrate nutrients. Scientists estimate that worm castings contain five times more nitrogen, seven times more available phosphorus, eleven times more potash, and forty percent more humus than usually is to be found in the top six inches of soil. For another thing, the castings mix the soil ingredients, facilitating their further breakdown by microbes. This has been proved in an ingenious experiment, ironically enough using carbon-14 laid down by successive nuclear tests as a measure.

The earthworm is also a pathfinder. It might be said that he weaves the soil with the thread of his leavings but also with the underground channels he creates. His blindness does not hinder his motion. A single acre of cultivated soil has been seen to have more than six million worm channels, whose presence significantly increases the soil's ability to hold and percolate water. A clayey orchard soil had more than two million large channels (some as thick as your little finger) in an acre, the equivalent, experimenters reported, of a two-inch drainage pipe. Other investigators have found that down to a depth of four inches, up to fifty percent of the soil's air capacity consists of the tunnels and cavities dug by worms.

Fishermen, who have a periodic and urgent need for the reclusive night crawler, have elaborated particular methods for coaxing him from the soil. Some say that they will come up if you pound on the ground, because they mistake your fists for rain. I bet. But the best way to catch worms is to "telephone" them.

The technique is named for the old-fashioned hand-crank generators that resemble old telephones. In fact, any generator or even a car battery will do. When an electrode is stuck into the ground, the night crawlers come directly to the surface. Why? "You would too," says Sybil Wallingham of the Wallingham Worm and Rabbit Farm in Butler, Georgia. "It's the same as if you're standing on wet concrete when the drill shorts out."

Soils, particularly when wet, are fairly good conductors of electricity. The worms' sensitivity to the charge is exquisite. In fact, they are markers of most calamities that happen in the soil. When a soil is too wet, the worms come to the surface. Just before an earthquake, too, they will often be found aboveground. Worms are the watchmen of the soil.

To encourage earthworms to live in your soil, provide them with plenty of organic matter. Professor Daniel Dindal of Syracuse University, an expert on soil organisms, suggests adding either a compost made from autumn leaves and manure, or freshly mulched leaves worked directly into the soil with a pitchfork and a hoe. Dindal also recommends adding a sprinkling of ground eggshells or other calcium-rich materials, since worms concentrate calcium in their guts and excrete it into the ground, where it can help to neutralize acid soils.

Is there nothing more arcane that needs to be done? "No," says Dindal, "just break up the materials so they are bite-sized for a worm."

There is no end of places where the gardener can buy worms, but there is seldom a need to, unless you are starting worm composting. Generally speaking, if you build soil, worms will come.

THE PHARMACY OF MOLDS

THE PROFESSOR OF SOIL SCIENCE SELMAN WAKSMAN was hoping that the rain would stop, since every time there was a good downpour, he and all his graduate students had to don galoshes to make their way around the basement laboratory at Rutgers University. Though all of them were students of the soil—working on humus, colloid chemistry, protozoa-bacteria symbiosis—it seemed more likely that they would write a new chapter of *Notes from the Underground* than any dry-eyed work of scientific inquiry.

It was after nine in the evening. Most of the investigators had stopped work for the day. With two bright and promising grad students from Europe, Waksman sat sipping coffee, trying to keep his feet dry.

The young Frenchman was having a hard time of it. His wife had contracted tuberculosis and was desperately sick. More than two centuries before, their countryman Manget had given the classical description of the effect of the disease upon the lungs. Reporting the autopsy of a young woman, he noted her lungs were "strewn with white bodies rather hard, of the size of millet seed, of the white poppy, and some of the size of a hemp seed, closely joined together, scarcely leaving any part of the lung free from them."

This was very nearly the condition of the young student's wife. It was virtually certain that she soon would die. The professor and the other student tried to comfort their colleague, but it was no good.

Just then, one of them asked a question that had been waiting almost a hundred years to be expressed. "Why is it," he said, "that when you bury a dead body in the ground, the earth is not poisoned?"

It was suddenly very quiet in the basement lab. None of them knew that they were asking the very same question that the poet Walt Whitman had asked three quarters of a century earlier in "This compost": "O how can it be that the ground itself does not sicken?" Neither was any of them yet in a position to discover the detailed truth of the statement. Yet it occurred to them in a flash that perhaps this was the question that they, as students of the soil, had been born to answer.

The corollary question was even more exciting: What diseases might the soil cure? Might it even destroy tuberculosis, the disease that prior to cancer was the greatest scourge? Could such a discovery save the Frenchman's wife?

In light of these questions, some of the folk remedies for tuberculosis came to seem far less crazy than previously. During the eighteenth century, the Methodist divine John Wesley recommended that the sufferer take cold baths, breathe into a hole cut into fresh earth, and "suck an healthy woman daily." Whatever use cold water and mother's milk may have been, it is not impossible that the soil would actually have been therapeutic. Another folk cure involved eating butter made from the milk of a cow that had grazed on churchyard earth.

But just where in the soil lay this healing virtue? Selman Waksman, the professor, and René Dubos, his French graduate student, set out to find the answer. On Waksman's recommendation, Dubos went to the Rockefeller Institute, where he succeeded in isolating from a soil bacterium the chemical that he named gramicidin. The first of the antibiotics, it was the basis for the discovery of penicillin.

But penicillin did not cure tuberculosis, much to the dismay of Dubos, whose wife had meanwhile died of the disease. Waksman, still working out of his damp lab, found the answer in another soil organism with the unpronounceable name *Streptomyces griseus.* The common creature—common as dirt, you might say—lives in many soils around the world, meaning that a person following Wesley's cure might actually have breathed in some of these. To destroy the tuberculosis organism, Waksman isolated the active chemical that streptomyces makes.

Streptomycin controlled tuberculosis for the first time. Along with penicillin and the whole stream of soil-derived antibiotics that followed—tetracycline, neomycin, etc.—it changed medicine forever.

Now for the first time, it was possible to use the power of one group of microbes against that of another, operating at the molecular level. The cures effected were little short of miraculous.

Dirt is the source of the greater part of our drugs against infectious diseases. And it may soon do for cancer what it did for tuberculosis. A certain kind of deep-living soil bacteria that exists without breathing oxygen creates in itself chemicals called enedeynes (een-DYE-ins). These complex molecules insinuate their way into the nuclei of cancer cells, where they cause an explosion that instantly ruptures the cell, killing it. No other drug yet found has shown such consistent and dramatic effects. There is now reason to hope that the soil may yet cure cancer.

However we succeed in manipulating it, though, dirt always outruns us. As Whitman wrote, "It gives us such fine materials and accepts such leavings from us at the end." Contrary to the scientific method, dirt works by ramifying possibilities. Instead of seeking a single solution to cure TB, for example, it elaborates hundreds of organisms capable of killing or arresting TB. The disease, meantime, organizes itself to resist the invader, limiting its losses to a minimum. In this way, a healthy ecological community can be maintained.

Molecular medicine intervenes directly in soil process, as well as in human growth. It cures human bodies before they have died and been buried, but it does so by means of soil organisms. It cannot, however, respond as fast to changes as the soil does.

The cycles of resistance in a living soil help to keep it in balance. If a bacterium is too badly depleted by another, it evolves new structures to withstand the threat. No creature is immune to competition, yet every creature is needed to keep up the web of reciprocal feeding.

This property, so important in ecosystems and in human physiology, means that nothing can permanently be cured. Nature is intent upon distributing her benefits to all her creatures, not only to man. If streptomycin attacks tuberculosis, someday a resistant TB will appear.

In 1991, new resistant forms of TB appeared in New York City, finding victims in a diverse population of drug users and AIDS sufferers. The currently available medicines are comparatively powerless against it.

As a result, there has been a scramble to find sources of streptomycin, which had over the decades been replaced by other soil-based drugs. The reemergence of TB and other infectious diseases has led the Rockefeller Institute scientist Sandra Handweger to declare, "It is time to go back to the soil." That is where new cures are to be found.

VII

Human Soil

THE SOIL AND THE DEVIL

THE DEVIL IS AN EAGER AND A SANGUINE ECONOMIST. He is the fountain of all plans to increase production by borrowing now and paying later. He has boundless optimism and faith in technology. His is the energy that drives a civilization upward until it topples under its own weight. He is not altogether a bad fellow, but he is most dangerous when ignored.

Just as all human wealth is tied ultimately to the produce of the soil, so the Prince of Darkness is intimately related to the farmer. In one of the world's great allegorical plays—Calderón's *auto sacramental, La vida es sueno (Life Is a Dream)*—the Prince and his partner, Shadow, become farmers in order to seduce Man.

It is not one of those mystery plays that plod on with the ponderous certainty of a dutiful drudge. Shadow wants Man for her own, to sleep with him. After all, she reasons, she was first created, before the Light, and she thinks it unjust that Power, Knowledge, and Love should decree that Man become the husband of Light. She *wants* him, and she will stop at nothing to get him. The Prince, guided by his Envy, is a willing conspirator. Anything that makes Man unhappy is delightful to him.

But the garden is fenced against them. If they come as they are, they will be recognized and expelled. At last, the Prince hits on a scheme: if they are disguised as farmers, none will question them in the garden.

When they appear, Nature is about to give Man her own fruits and flowers as the tribute due to him as Lord of Creation. Shadow pushes her aside, saying:

"Get back, you rusticated earth! All you'd give him
were little wildflowers, were it not for the industry
by which I add to your crude forms
all the paints that make them shine.
You give birth but I refine,
and so the fruits are mine."

Man falls in love. The apple that Shadow offers him is obviously bigger, better, and tastier than any he has yet experienced. "Who are you, lovely maiden," he breathes, "mistress over all Earth's plenty, who bears away the prizes that Earth cannot keep from you?"

The apple is totally free, Shadow tells him, and when you taste it you will become just like God. But when he does bite into her gift, the Four Elements, until then his servants, depart from his side. The garden draws away from him and he is left to sweat, farm, and die with Shadow.

Ever since that primal scene, we have no excuse not to know the devil and his accomplices, but the farm has been the perpetual theater of his conquests. In one respect, we wish to be taken by him, for he is Eros and he moves us to increase. In some respects, every pregnancy and every crop is a bet on the future, a speculation that somehow resources will be found. As John Adams said to Thomas Jefferson, if we didn't have desire, no one would go to the trouble of having children.

In his other aspect, however, we wish the devil would go away, because he is chaos and collapse. The Greeks had the good sense to split off Eros from Hades, making them two principles, but as messy as the devil is, his dual nature creates a useful heat and friction. Indeed, the rationalist divines of the eighteenth century so feared the energy of Calderón's Prince that they banned the allegorical plays in which he is always a leading character. As Blake wrote of Milton, "He was of the devil's part without knowing it."

Calderón and Milton both kept the devil visible. They told the whole story by bringing darkness into conflict with the light. The rationalists put the devil under taboo and so gave him free rein to operate unseen in obscurity.

The soil, in its darkness, is under the special protection of the devil. It is the source of all generation of life, and it is the place to which the dead return. Hell is always underground. When Yahweh kicked Adam and Eve out of the garden, he specifically made the soil "accursed," that is, he endowed it with this double principle of good and evil, rise and fall. To forget the devil's doubleness when dealing with the soil is a dangerous thing indeed, because instead of harnessing his energy, we leave free his power for crime.

Every time the progressives are in control—I mean, those people who tell us that everything is gradually getting better and better, and that if a certain technical problem has not yet been solved it soon will be—you can wager that the unseen devil is rummaging about in the soil wrecking havoc, while his bright side, the sanguine economist, is predicting a painless era of plenty and heaven on Earth.

This is not a new phenomenon, caused by one economic system or another, but a perennial struggle. The first cities of the Western world rose around 4000 B.C. in the fertile triangle between the Tigris and Euphrates (which, by the way, are two of the four rivers mentioned as running through the Garden of Eden). Their wealth was based on the first state-organized farming. The Mesopotamian cultures were the first to introduce the plow, the yoke, the potter's wheel, and, above all, large-scale irrigation. Growing primitive wheats and barleys, the Sumerians, Akkadians, and Babylonians cultivated thousands of acres adjacent to the rivers. On the basis of these, grew their ziggurats and warring city-states.

To look today at those barren lands, dotted here and there with hillocks that once were proud temples, you would never think that Western civilization was sourced there. But all of these early cultures were defeated by a devil called Tiamat. Her weapon was salt.

Water carries dissolved salts. If these are not ultimately deposited in the sea, they concentrate in the groundwater. The Tigris and the Euphrates are very long rivers that deposit most of their salt load before they reach the sea. (The Nile, by contrast, has supported agriculture for millennia, because it flushes its salts with an annual flood.) They also leave layers of silt on their bottoms, gradually elevating the course of the rivers.

The Mesopotamians increased tenfold and more the total production ever achieved before irrigation, and with the help of their gods, they expected to prosper limitlessly. But Tiamat was at work unseen in the soil. In the irrigated fields, the water table rose, not only because the siltation was elevating the river bottom, causing the rivers themselves to rise, but also since irrigation brought water in far more quickly than percolation could dispose of it. For a few centuries, though, everyone prospered.

Then, the flowers began to appear. White, tinged with red and yellow, they spread across the flat surface of bare soil. They were deadly flowers of salt. At first, the farmers resorted to barley, a more salt-tolerant crop than wheat. Then, they practiced fallowing to allow the water table to drop between plantings. But eventually field after field had to be abandoned.

The farmers did not give up without a fight. Even when populations had tripled, and yields per acre had declined to only one fifth or even one tenth of what the fields had first yielded centuries before, they struggled to feed the cities. The only way to do that, however, was to double the rate of seeding and to use the fields intensively, never resting them. So the harder they tried, the more grain they wasted, the quicker the salt rose, and the more yields declined. Tiamat won. By the year 1000 B.C., most of southern Babylonia had been completely abandoned.

Ignorance of the devil always leads down this path. More technology, greater planting rates, more intensive use, greater dependence on larger holdings, and fewer farmers are supposed to save the day. Instead, they hasten decline. The Romans found out about it when their immense slave-farmed estates were exploited beyond the soil's tolerance, leading to massive erosion and declining fertility. Medieval Europeans became so desperate to feed their booming cities that they took the leaf mould from the forests to use as compost on the fields, improving the latter briefly at the cost of the destruction of the former. Forests declined not simply from overcutting for firewood and ship's timber, but because the soils beneath them were robbed.

History's great observer of this process was William Cobbett, whose *Rural Rides* documents the destruction of the English countryside during the nineteenth century. He set clearly the distinction between the yeoman farmer—who owns, lives upon, and therefore knows the land—and the absentee landlord who rents it to cultivators whose only job is to extract from it as much profit as possible.

The economist runs roughshod over the health of the land. Of this Cobbett had no doubt. "It is the destructive, the murderous paper [money] system," he wrote, "that has transferred the fruit of the labor, and the people along with it, from the different parts of the country to the neighborhood of the all-devouring Wen." The word "Wen," meaning literally a boil or carbuncle on the skin, was his name for London.

The national debt, taxation, and the stock markets were in his view inimical to the soil's health, because they tended to concentrate wealth into great, compact masses. These concentrations benefited the few who owned them, and benefited the governments, because they provided a convenient and available source of tax revenue. But great parcels of land could not be treated like stock certificates, or when they were, the result was erosion, poverty, and declining yields, just as the increasing populations of the city called out for more food.

But the devil is nothing if not resourceful. No matter how often the same situation recurs and how often it is pointed out, the economists still stand up to claim that everything is getting better and better. The devil's most charming and pernicious argument is the argument from history itself. He says, "Well, all the past problems have been solved by human ingenuity, so the current ones will be too." He does not factor into the equation the vast human misery that has been created along the way, for this suffering, after all, is his deepest delight.

THE SOIL APOCALYPSE OF GEORGE PERKINS MARSH

GEORGE PERKINS MARSH FOUNDED AMERICAN ECOLOGY with his 1864 *Man and Nature,* a study of the Earth as modified by human action. Here is his vision of the nineteenth-century destruction of a landscape:

> *With the disappearance of the forest, all is changed. . . . The face of the earth is no longer a sponge, but a dust heap, and the floods which the waters of the sky pour over it hurry swiftly along the slopes, carrying in suspension vast quantities of earthly particles which increase the abrading power and mechanical force of the current, and augmented by the sand and gravel of falling banks, fill the beds of the streams, divert them into new channels and obstruct their outlets. . . . From these causes, there is constant degradation of the uplands, and a consequent elevation of the beds of water-courses and of lakes by the deposition of the mineral and vegetable matter carried down by the waters. . . .*
>
> *The washing of soil from the mountains leaves bare ridges of sterile rock, and the rich organic mould which covered them, now swept down into the dank low grounds, promotes a luxuriance of aquatic vegetation that breeds fever and more insidious forms of mortal disease, by its decay and thus the earth is rendered no longer fit for the habitation of man.*

HUSBANDRY IN ROME

VIRGIL HAS ALWAYS BEEN IN MY TOP-FIVE HEROES LIST, right behind Robin Hood. He wasn't so good with pike or bow, but he got Dante through Hell and most of Purgatory.

Midway through Purgatory, Dante and Virgil meet the swaggering Italian poet Sordello. "So who are you?" Sordello asks of the tall shade that walks beside Dante.

Virgil responds by saying that he is the poet whose funeral was presided over by Octavian Caesar, Emperor of Rome. It dawns on Sordello just who he is talking to, and the Italian falls on his knees.

"O glory of the Latins!" he exclaims, as all poets have since then. Virgil was author of the *Aeneid*, the epic of the founding of Rome. But he wrote another epic too, one that was equally famous during his lifetime. It is the *Georgics*, a long poem about dirt.

The *Georgics* were composed at a crucial moment (36–29 B.C.) in Roman history. Julius Caesar was dead. Marc Antony bid fair to fill the power vacuum, while the young Octavian Caesar waited in the wings. Then, with the suddenness of a single battle, Octavian was in power, the Roman Republic was a thing of the past, and Rome was embarked upon its era of greatest influence and affluence.

The usual modern idea about the *Georgics* is that Virgil wrote them to look back in longing toward a simpler time. The turmoil of the incipient empire, according to this view, made for a market in nostalgia. This is rather like calling Virgil a back-to-the-lander, and it sounds suspiciously like the arguments that dismiss Jeffersonian agrarianism as a backward-looking romantic ideology, irrelevant to our times.

Yet perhaps we are the ones who have it backward. Perhaps it is our time that looks through the wrong end of the telescope, and so can scarcely make out a large truth that Virgil, his patrons, and readers—and

Jefferson, too—knew well. Nobody needed the *information* that Virgil put into the *Georgics*; indeed, he took a fair amount of it from his contemporary Varro, whose *De re rustica* (*On Farming*) was the best agricultural manual of its time. Virgil wrote the book for another reason.

Husbandry is the art of daily life, of observation and response. It is not simply a list of practices, but an attitude toward living that entails honesty and economy. In the writing of it, Virgil is the poet of a whole way of life. Among the finest descriptions in Latin literature is that with which he begins the book:

When Spring is new, and frozen moisture thaws
On white-clothed mountainsides, and crumbling soil
Is loosened by the West Wind, let your bull
Begin to groan beneath the pressing plough
And the well-worn ploughshare gleam from the rub of the furrow.

Economical and deeply erotic, the description brings the reader from the cosmic to the local in the space of less than fifty words, a precis of the book. Through the succeeding lines, Virgil tells the farmer how to match his wishes to the "native traits and habits of [each] place."

How different is this relationship to place from the hysterical and hurrying ways in which we typically regard our yards, farms, and homes! A rhythmic and measured relationship to time is one of the prime virtues that the *Georgics* note. The husbandman is also for Virgil a type of the just man.

The farmer lives in peace, his children all
Learn how to work, respect frugality,
Venerate their fathers and the gods:
Surely, Justice, as she left the earth,
In parting left her final traces here.

Husbandry, to Virgil, is a question of right scale and right relationship, an economy bound by trust as much as by fear. And he is under few illusions that it is easy to practice. The *Georgics* are a challenge thrown in the face of his times.

But we think we know better. We know that a patron of Virgil's sold his business and moved to the country after reading one of the pastoral eclogues (a poem in which two shepherds converse), only to regret it quickly and return to the bank. We know, too, Flaubert's delicious pillorying, in *Bouvard and Pecuchet,* of the petit bourgeois pair whose efforts to return to the land are comically thwarted.

We think we know better, but we do not take into account all those who have indeed established a new relationship to the land. Robert Frost was one. Another was the writer Louis Bromfield, who, having conquered Broadway, Hollywood, and Europe, returned to the Ohio farm valley of his childhood and wrote three classics about the farm he founded there. At the end of *Pleasant Valley,* the first of these books, he thanks the land for giving him "the richest and fullest life that I have ever known."

The cynic says, "So what! Did these guys make a living off those farms? No, they were writers, so they didn't have to."

But what if their first thought were not simply to "make it pay," but to be paid in the coin of food for honest work? Work is itself a form of pay. We are now entering a time when there is so little real work left, that people will pay a great deal to be allowed to do it. I think of all the archaeological digs, the cattle drives, and tree plantings that people will pay for the privilege of participating in. What is prosperity anyway? Is it money in the bank, or is it a fullfilling life?

In the heyday of the Roman Republic, around 150 B.C., things were not so different from things now. Cato the Elder, the first of the great Roman agricultural writers, characterized economic life in this way:

> *The pursuits of commerce would be as admirable as they are profitable if they were not subject to so great risks: and so likewise, of banking, if it was always honestly conducted. For our ancestors considered, and so ordained in their laws, that, while the thief should be cast in double damages, the usurer should make four-fold restitution. From this we may judge how much less desirable a citizen they esteemed the*

banker than the thief. When they sought to commend an honest man, they termed him good husbandman, good farmer. This they rated the superlative of praise.

An honest relationship to work requires a human scale. The multiplications of potential occasioned by banking make possible vaster-scale enterprises, but run the risk of turning wish to greed. Already at the time of Cato, finance was beginning to take its toll on agriculture, driving individual farmers off the land and instituting a system that resulted in ever-larger farms run either by tenants or, increasingly, slaves.

Virgil counseled, "Admire a large estate, but work a small one." With the expansion of Roman territory and the need to defend it, however, the countryside gradually became depopulated. A large estate became the norm, a small farm the rare exception.

Already in his time, the slave-worked farm was common. Rome had lost over 300,000 men in the Punic Wars, most of them farmers. Consequently, the slaves who had been brought in to replace them during the conflict remained when the war was over. The land passed into the hands of absentee landlords, a gentry who themselves lived in the cities.

By the time of the first-century writer Columella, Rome had a vast urban infrastructure and a population dependent upon the produce of slave-run farms operating on the margins of the empire. In the cities, the masses lived on handouts of bread made from wheat from these farms. Columella, though he extolled the virtues of the independent farmer, recognized the influence of both the banker and the lawyer in destroying farms. The lawyer's profession, he reports, was called a canine one, because it consisted in barking at the doors of rich men.

Meanwhile, the far-flung empire needed more and more defending. Taxes increased, conscription was general. Though the government made it possible with one hand to keep slaves on the farm by offering proprietors the right to buy them out of military service, on the other hand, the government imposed such crushing taxes on the produce of the fields that the entire harvest might scarcely repay the cost of growing it.

Farmers, caught between a shortage of labor and uneconomic crops, began to abandon the land. As they did so, it was left bare, open to the rains. The earth eroded, and the fields were destroyed. In this respect, Rome fell not under the pressure of the barbarians, but from her own measureless girth, which destroyed her relationship to daily life.

Recently, I asked the writer and farmer Wendell Berry why he continued to work the marginal land of his farm on the banks of the Kentucky River. As he himself said, probably no one else would choose to farm it after him. And he didn't make a profit on the farm, though its produce dependably supplied his family with much of their subsistence.

His children are grown and gone. On a rainy day, it looks almost as though the farm were going to slip into the muddy brown river. Why keep doing it? "For forty years," he said, "this farm has been the best education that I ever had."

DIO-HE-KO

Dio-He-Ko
Dio-He-Ko
Corn Beans Squash
Corn Beans Squash

SO SANG THE MAIDENS of an Iroquois agricultural society. They worshiped a three-in-one goddess who appeared in the world under the guise of corn, beans, and squash, the three crops that were planted together in Iroquois village plots.

What were the virtues of this goddess? She was herself a community. As *Dio,* corn, she grew tall and sustained the people with her fruit. (A native farmer could sing the song that counted every leaf, every joint, every tassel in the order that it appeared on the plant from seedtime to harvest.) As *He,* beans, she twined about the corn stalk, making of it a living trellis on which to support the vine and its fruits. In return, as beans, she fed the roots of the corn and the squash. And last, as *Ko,* squash, she put out spiny leaves and stems to defend the plot from marauders.

This triple goddess—a kind of New World Hecate—is the founder of the Thanksgiving feast, and so it might be said she is ruling deity not only of the agricultural native peoples but also of the newcomers who survived their first winter only owing to her bounty. It is not just bad policy to reject her, as we have done over the past three centuries. It is sacrilege, and it is being avenged every day in farm fields across the continent.

I talked to an agronomist who'd tried the Dio-He-Ko system in Vermont during the 1980s, only to find that the raccoons were by no means sympathetic to this poetic association, devouring both corn and squash without mercy. To the cynical, this would be prima facie evidence that the Native Americans as farmers were not all they are cracked up to be. But remember that in the first place, raccoons were not so dependent then as they are now upon the leavings of man, nor were they hemmed in by roadways, nor had they multiplied beyond the capacity of the ecosystem to maintain them owing to the destruction of the larger mammals that preyed on them. Furthermore, it is important to remember that Native American horticulture was based on the principle of sharing, so that a portion of the produce was by right set aside for the likes of coons, woodchucks, and deer.

"Polyculture" is the name for this way of growing useful plants in associated groups. And it is not the pipe dream of backward-looking romantics. In fact, most nonindustrial agriculture even today is polycultural. Contrary to expectation, the difficulty that Dio-He-Ko experiences in our American landscapes is probably more the result of her degradation than of the impracticality of the idea.

Polyculture in the tropical Americas thrives wherever plantation or intensive systems of export agriculture have not been substituted. In the 1950s, the botanist Edgar Anderson studied a Honduran plot the size of a New York City backyard where a single family grew more than thirty different crops. Some were trees grown for their fruits or fiber, some were understory shrubs with properties useful in medicine rituals or bearing edible nuts or fruits, and some were herbaceous plants whose leaves, seeds, or tubers formed the staple of their diet. No fewer than three hundred species of animals—fifty of them ants alone!—cohabited on the same plot, and no one went hungry.

The same mixed culture is characteristic of the traditional slash-and-burn agricultures of the Amazon Basin. Burning out a section of forest, the cultivators would take advantage of the nutrient-rich ash to start crops, including seventeen species of manioc, all on the same acreage. As the plants exhausted the nutrients—aided by the torrential rains that sluice through the porous soils, carrying away anything soluble—the yields would decline. The farmers responded by ceasing to make new plantings.

Instead, in an act of real and millennial symbiosis, they would harvest both the survivors of the plants and the successor species that established themselves in the now-uncultivated gaps. Who knows what give-and-take over millennia might have resulted in this situation, where even as the farmer gives back the forest to itself, she is permitted to realize more use from its products. Eventually, the forest would close the gap, but by that time, the farmer would have moved on to a new plot, burning it off and beginning again.

The superiority of such a system, both spiritually and materially, over the monocrops of cattle pasturage that recently have been installed in these regions is obvious. To maintain a single species over time on the cleared ground requires constant inputs of fertilizer, in addition to the ash of burned trees. Without the fertilizer, the hungry pasture grasses won't thrive. Even where the fertilizer is applied, however, the rate of leaching is often so fast that the grasses grow with less than their accustomed vigor. The cattle who eat them are therefore undernourished, particularly with respect to phosphorus. As a result, they break bones when they stumble.

The difference between the harmonious web of the polyculture and the Rube Goldberg mishaps of the monoculture is palpable, but one might well object that such a polyculture is impossible wherever land is not abundant. After all, the system does depend upon the ability to abandon one plot at will and quickly find another (at little or no cost) to replace it. It is not impossible to return to the original plot, but not for at least a decade.

Where populations are larger and land is a commodity, it seems clear that polyculture is not an option, unless some means can be found to replenish the soil in situ, without the humans' departing for ten years. The native populations of New England, where the soils were poor and stony to start with, solved this problem in two ways. First, in the Dio-He-Ko group was a legume, beans. They knew nothing whatever about the molecular chemistry of nutrition or about the fixing of nitrogen by legumes, but their observations told them all they needed to know: A plot with beans grows stronger than one without. Furthermore, to provide corn, the heaviest feeder, with sufficient nutrients, they knew to throw a menhaden or another small fish into the planting hole.

The colonists, who grew the crops not only for subsistence but for exchange value, truncated this system and so destroyed it. The picture of the devastation, as William Cronon describes it in *Changes in the Land,* is so frightening that it makes one wonder how people could go on behaving in just the same way for the nearly half a millennium since they destroyed New England farming.

Corn was the cash crop, the colonists perceived, so they grew it to the exclusion of beans and squash, which were relegated to separate, smaller monocultures. Nevertheless, they had observed that native success with corn was owed to the habit of applying a fish. In grotesque imitation of this practice, they fished out entire rivers, applying the stinking carcasses by the thousands to their cornfields. Country travelers spoke of an "almost intolerable fetor," which, given the already powerful odor of the average colonial gentleman, must have been foetid indeed.

The results were fished-out rivers and dying soils. Efforts to revive the fields with wood ash resulted in deforestation. By the time someone hit upon the idea of using clover to regenerate the soils, many fields were already far gone. Today, from New Jersey to Maine, there are thousands of miles of stone walls that once marked the boundaries of farmers' fields. Now they are hidden among forest trees. In the long run, the monocrop failed, the forest won.

Yet in the American experiment with massive monoculture, it was only a temporary setback. Farther west was fresh land that had been covered with native prairie for at least ten thousand years. It would take a long time to destroy those fertile soils! And so it has. A little more than a century. Only now are the modern equivalents of the New England abuser beginning to test the limits of the tolerance of the great brown mollisols of the Midwest.

Polyculture here is a laughable notion, though the fertility of the soils is owed largely to their ancient prairie cover. Mature prairie is among the most extreme herbaceous polycultures imaginable, 150 or more species growing together in a single association. For this, farmers substituted huge fields of single crops—corn, soy, or wheat—and stripped even these from the land during the winters, when wind and rain would erode and gully the soil.

This disastrous situation came near to ending American agriculture during the early 1930s. It was then, however, that the desperate remembered the old lessons of polyculture. The 1938 Department of Agriculture yearbook, *Soils and Men*, one of the great practical treatises ever published in this country, gave detailed and experimentally verified methods for restoring the soil by imitating polyculture. Indeed, this is what crop rotation and manure use amount to. In the polyculture, the green manure crop grows among the other crops at the same time, and the animals who share the meal pay for it with their droppings. On the farm field, the clover crop is separated from the corn crop by an interval of time, and the cows may be separated by miles of distance, but the principle is the same: return what you remove by the means that built the fertile soil to begin with.

These methods were well known to such venerable agricultural systems as the Roman and the Chinese. Perhaps the great statement of them comes from Virgil's *Georgics*.

. . . when the seasons shift
Sow in the golden grain where previously
You raised a crop of beans that gaily shook
Within their pods, or a tiny brood of vetch,
Or the slender stems and rustling undergrowth
Of bitter lupine. Crops of flax burn out
A field, oats burn it through, and drowsy poppies
Soaked in oblivious sleep will burn it too:
But still rotation makes you labor easy,
As long as you are not ashamed to drench
The arid soil with fertile dung. . . .
Thus will the land find rest in its change of crop,
And earth left unploughed show you gratitude.

The ideas reappeared in twentieth-century America just long enough for chemists to invent another grotesque shortcut. By learning to fix nitrogen from air, in a high-temperature industrial process, they expected to do away with the messy and inefficient use of cover crops and manures. That is just what they have done, and it is hard to believe that the results will not be at least as unfavorable as those in colonial New England.

Perhaps, however, there is a way out. A farmer might intentionally return his farm to an actual polyculture of perennial plants with more or less permanent roots in the soils. This was another system adopted both in Rome and its provinces and in premodern China: interplanted with and surrounding fields of herbaceous crops were to be found groves of olive or mulberry trees, and vineyards.

The trouble is, however, that Midwestern soils are not usually supplied with a tree cover, and virtually all crop plants that form our dietary staples are annuals, which set seed and die, roots and all.

Enter geneticist Wes Jackson, with an idea as powerful as it is bold. If we do not have any perennial crop plants, Jackson proposes to breed them. Never mind that the whole course of plant evolution militates against a good seed bearer also having perennial characteristics. Jackson exudes confidence. "In a generation or so we'll have the problem solved," he says. Having seen the limits of rotation and of manuring, he argues that we take natural systems as our standard, in order to regenerate the soils. Though trained as a geneticist, Jackson is a deep and reverent economist. He envisions farms that "run on sunlight"—making maximum use of energy derived from sunlight and minimum use of fossil hydrocarbons like oil and gas.

It will take more people on the land, more shared work, and more community to accomplish these ends, he believes. And though he has no wish to reestablish the Dio-He-Ko model itself, his vision of a new polyculture offers hope not only for agriculture but for the culture of towns and communities.

VIII

Visions of the Soil

UNDERGROUND HORIZONS

HANS JENNY USED TO SAY that soil is a Body in Nature. I wondered what he meant.

To try to understand the soil by taking a few trowelsful and submitting them to chemical tests is like trying to understand the human body by cutting off the finger, grinding it to paste, and performing the same tests. You may learn a lot about the chemistry of pastes, but about the intricate anatomical linkage of systems—and about the body's functions as a whole—you will learn nothing at all.

Like our bodies, the soil participates in the recirculation and transformation of the four major elements: earth, air, fire, and water. Like our bodies, too, it is full of channels and pathways, directing the elements into fertile combinations and transformations at distinct, organized levels of the whole structure. And like our bodies, it has a definite genetic form.

Jenny once described a dog digging dirt out of a hole. "That," he said, pointing to the mound of earth, "is not a soil." To be more precise, he added, "Or it has only begun to be a soil." In essence, it is no different from the ash just spewed out of Mount St. Helens. Its particles are more or less isomorphically arranged, randomly ordered. Its pebbles are still weathering to spheres, the shape of least resistance, not reorganizing and building energy at the microscopic level of the clays and iron oxides, as a soil does. Its tissues have not become differentiated. But the moment it is exposed on the Earth's surface it begins to acquire a body, to incarnate, as it were, to become a soil.

The body of a soil is a sky where seeds and worms and ions fly. Just as the sky links outer space to Earth's surface by means of increasingly dense atmospheric layers, so the soil links the surface to planetary bedrock by means of increasingly dense layers called, appropriately, horizons. Where the bottom layer of the sky rubs up against the top horizon of the soil, all terrestrial life is found.

In a fresh roadcut or beside a beach, you often can see the soil's body exposed. It can be very beautiful. One February morning, I slammed on the brakes a few miles from the Tallahassee Airport. Cars zipped by. Drivers looked at me in anger or puzzlement. But I had found a very sexy soil-cut maybe twenty feet deep, created by the highway department as a source for fill to line the roadside ditches that eroded every spring.

Atop it was a forest of small second-growth pine, but the dramatic thing was the soil. Its upper horizon, maybe six inches deep, was dark with the decayed remains of pine needles, bark, and the root and stems of grasses. Beneath that was a sandy layer taller than me, the color of bone china or beach sand, looking like a Sahara in the making. And beneath that, stretching another ten feet down to the base of the cut, and who knows how far into the underground, was an orange-red horizon, variegated like a sunset on the ocean.

How is this picture different from, say, the lovely strata of Grand Canyon sandstones? Geological strata are the product of differing climates and different influences, separated by large periods of time. They express a calendar or an almanac. The horizons of a soil, on the other hand, express a single life, rising out of a single environment.

The soil therefore has an age, an expression of the life it has lived in its own peculiar place. In Tallahassee, where temperatures and rainfall are both high, and where the parent material was marine sand left high and dry after the end of the surge in sea level when last the glaciers melted, a soil on level ground very quickly becomes deep.

Rainwater, made more acid and chemically active by picking up CO_2 in the air and by the residue of the pine trees, sluices through the sand at a rate of twenty inches per hour, or even higher. It might as well be going down a bathtub drain, except that here the water chemically extracts and moves aluminum and especially iron, carrying it down into the subsoil, which it makes red, leaving the unmoved silica above white. Within a few thousand years, the soil is fifty feet deep, and the native slash pines are sending their taproots down to find water.

The cut was a place of surprises. Not only were some tree roots exposed, showing a taproot more than three times as deep as the sapling was tall, the red subsoil was also the scene of unexpected developments. Nothing at all was growing in the exposed whitish horizon; for all the life evident in it, we might have been flying in the clouds. But beneath it, in the red layer, streaks of green and purplish lichen were colonizing rough runnels, and here and there, the tiny green bottle brushes of the club mosses had begun to emerge. The exposed subsoil, like the dirt the dog dug up, was already beginning to become a new soil in itself, with its own biota. In the structure of the old soil were contained a new heaven and new earth.

Horizons are what make some people become soil scientists. They are that lovely. To describe the variety of soils and their profiles, the thinkers have come up with a whole set of strange names that they nevertheless pronounce lovingly, like a best friend's nickname. Joe Shuster, of the Soil Conservation Service in Tallahassee, helped me wonder about the name of the soil I'd seen. "It sounds a little like a haploquod called Leon," he reflected. "That's close to a Myakka soil, the state soil, of course." (It was news to me that there was a state soil, but Shuster didn't miss a beat.) I described the odd little club mosses growing in the subsoil. "Sandy, was it?" asked Shuster. Yes, I said, it was. "And kind of dry?" Yes. "And what was the relief?" he asked, warming to the game. I told him it was flattish.

"No, no, I take back that Leon," he concluded. "I'll bet it was a Mandarin soil, that's a haplahumod, a little bit drier than a haploquod." I told him thanks, that's exactly what I needed to know, but would he mind spelling the terms?

While he was doing that, he evidently had time to consider. "You know," he added, with a tone of awe in his voice. "You almost saw a Resota." I checked myself before asking what kind of hapla that was, for fear it might be something else even more difficult to spell, like an argialboll, an ustochrept, or a quartzipsamment. Dr. Shuster, however, needed no urging.

"R-E-S-O-T-A," he spelled. "It has a much thicker spodic horizon, and there are long tongues of the Albic horizon that extend down into the yellow Bw horizon. It's very impressive." He paused to let that salvo take effect, then added, "And you should see the spodic horizon. It's intermittent!"

I couldn't flip the pages in my nomenclature book fast enough to figure out just what he was talking about, but his tone made me feel like hitting the road in search of the great Resota. I kicked myself for having missed it, even as Dr. Shuster reassured me that my Mandarin was *almost* as spectacular.

It's easy to laugh at a nomenclature so arcane that it includes ten soil orders, fourteen thousand soils with proper names, and twenty-one letter designations to distinguish the different characteristics of soil horizons. Yet from another point of view, this naming by kinship admits life to this matter. In being related to each other, they are at another level related to us.

The assembly of horizons, the profile, is the signature of any soil. And the profile is created by the dynamic interactions of the four elements. Earth is present in the parent material, the rocks whose weathering gives the soil its bulk. Usually, earth is passive in the reactions that make soil, being transformed by what reaches it. Water is the most obviously active matter, the agent by which silica, irons, aluminums, clays, and humus are driven down from the surface into the subsoil, creating horizons like my Mandarin's white and red ones.

Air and fire are also active. In a soil that has lime in the parent material, like the fertile soils of the Midwestern prairie, carbon dioxide is a prime mover in the creation of the soil. It engages in a dance of solid and liquid phases, whose result is the storage of lime in a white horizon in the subsoil. The roots breathe CO_2 out into the soil, as do the millions of soil microbes and fauna as they digest organic matter. That gas changes ordinary calcite, the comparatively immobile calcium carbonate, into calcium bicarbonate, a form that dissolves readily in rainwater. The dissolved matter then sinks through the profile, until it reaches a depth at which there is no longer sufficient carbon dioxide to keep the bicarbonate reaction going. The matter reverts to calcium carbonate and falls from solution, creating a white horizon of calcium, like a layer of cirrus clouds.

A hotter soil goes deeper faster. The rate of chemical reactions doubles for every ten-degree Celsius rise in temperature, transforming primary minerals—silica and aluminum into clay minerals, and iron into oxidized iron compounds—and freeing salts, acids, and bases to further work on the soil profile. Another dance of matter takes place. Dispersed and stable clays, in the presence of electrically energetic compounds, curdle into heavy lumps. Gravity drags these lumps deeper into the profile, until the decrease of salts allows the clay particles to disperse once more, forming a clay layer in the subsoil.

By the same token, when chemical reactions release hydrogen ions into the soil, it becomes more acid. The more acid in the soil solution, the more iron compounds become soluble. In fact when the pH (a measure of hydrogen ion concentration) goes from 8.5 to 6—a logarithmic increase of 250 percent—iron becomes 100,000 times more liable to dissolve. The rainwater washes it deeper into the profile, where it forms a rust-red horizon at the place where the hydrogen concentration drops.

A soil is not a pile of dirt. It is a transformer, a body that organizes raw materials into tissues. These are the tissues that become mother to all organic life.

WIND AND SOIL

DUST IS OUR SOURCE OF FERTILITY. The glaciers and the deserts feed the stable platforms, or cratons, like those broad prairies on which I'd found a former sea. From the frozen and the arid wastes come the materials for luxuriant growth.

The surface of the Earth is a thin film that exists between two boiling systems. Deep beneath the ground, the roiling of the hot mantle propels the crust's tectonic plates. Above the ground, the winds of the world execute the same ballet, girdling the planet from east to west. The winds ascend and descend in the spirals of convection cells, rising when warmed and sinking as they cool. If you could see these currents, it would look as though the Earth were wearing a knitted sweater of horizontal bands.

When the winds are descending, they gather moisture and heat energy. Little rain falls beneath them. When they ascend, the wet air cools, clouds form, and it rains, sometimes copiously. The Sahara and the Gobi Desert are located where the wind is descending; the rainforests of the equatorial zone occur where it is ascending.

Where the ground is dry and the soil is fine, a descending convection cell can start a dust storm, the same thing that happens when you come down fast on dust bunnies with a broom. The cooled air spreads across the ground like a splash, whipping up the silt into turbulent eddies. The dust drifts into the upward-bound part of the cell and quickly finds itself ten thousand feet in the air. Such storms, traveling at speeds of up to 650 feet per second, can drop the dust hundreds or even thousands of miles away.

This is what happened in the periods of glacial retreat that punctuated the Pleistocene Age, the last one beginning some fifteen thousand years ago. When the glaciers drew back and the climate warmed, the former ice sheets released tons of rock flour from their masses, leaving it along the watercourses of the outwash plains. Then, as the climate dried and cooled, the silts rose with the winds and were scattered over the upper Midwest and down the course of the Mississippi River Valley.

The dust storms that raged then would have made the 1930s' Dust Bowl look like a wheeze. Landscapes that started out relatively naked were covered with porous silty soils more than a hundred feet deep within a few centuries. It is still possible to trace the course of these winds over the land, because as they traveled they dropped the heaviest particles nearest to the source, the finer particles remaining longer in suspension before they finally settled to form the best of the prairie soils. These fine soils were both rich in nutrients and porous in structure, so that water did not pool in them, and so that nitrogen, potassium, and phosphorus quickly became available for plants. Once the prairies were broken for corn, these soils yielded three times the harvest of nearby claypan soils.

The American prairie soils are not unique. The Russian chernozem, the amazingly productive black soils studied by V. V. Dokuchaev, the pioneer of soil science, also have their origin in this windblown dust called "loess." The deepest soils in the world, the yellow loams of northern China, some of them hundreds of feet in depth, from whose sediments the Yellow River takes its name, were formed by dusts that are still being blown into the region from the Gobi Desert. Some of the people of this region traditionally excavate their houses directly into the material of these dirt cliffs.

So the Dust Bowl of the American Midwest in the 1930s was not an aberration, but a miniaturized replay of the forces that caused the region's fertility in the first place. Unfortunately, it worked in reverse. The rains failed, and the cultivated soil became once more the mobile dust that it had begun as. A writer for the U.S. Department of Agriculture put the matter movingly in 1938:

Everything goes well as long as the rains continue. But inevitably the drought years come, as they have with dismaying regularity since 1930. The crops fail and the pastures and range grasses stop growing. Stock-water reservoirs dry up, and cattle are rushed to market to prevent them from starving or dying of thirst. The dust begins to blow, and black blizzards lay bare the soil down to the furrow bottoms, pile drifts of dust around the farmstead and in the fencerows, and make life unbearable for all the people within a radius of several hundred miles.

When a dust storm rose, the sun turned red and the day darkened to dusk. As slippery and small as its particles are, the silt got in everywhere, even under sills stuffed with towels. People caught in the storms sometimes choked to death on the silt.

In 1937, a dust storm started out in the Texas and Oklahoma panhandle country, crossed five states, and dropped the last of its load in Canada. As the storm churned, it picked up much of the fine soil at the source, leaving only coarse sand behind. As it traveled, it dropped the coarsest materials first, leaving the finest only at its last stop in Canada. When tested five hundred miles from its source, the storm's dust was found to have ten times as much organic matter, nine times as much nitrogen, and nineteen times as much phosphoric acid as did the dune sands that were left at the storm's source.

Five hundred miles is piddling. The fact is that the Brazilian rainforest is an artifact of the Sahara, some thousands of miles across the Atlantic. The silt originating near Lake Chad in the southern Sahel is what fertilizes the rainforest. Great plumes of dust rise from the desert and drift west, dropping their coarser particles in North Africa. As they pass out over the Atlantic, they meet the edges of the storms that will soon pelt the Amazon Basin. University of Virginia researchers estimate that twelve million tons of Sahara dust drop on Amazonia every year, bringing to the nutrient-poor soils a shot of fertility. Particularly important is the available phosphate, about a pound per acre, that the storms bring; there is virtually no other available phosphate in the old, deeply leached Amazon soils.

The scientists were not even looking for dust when they learned this startling fact. They were part of a 250-man international team funded by NASA and its Brazilian equivalent to learn more about how the hydrocarbons released in Amazon burning might be affecting global warming. "We were looking for hydrocarbons," recalls Robert Swap. "When we started finding soil particles in the air, it blew us away."

At first, they thought that the dust must have come from the arid lands in northeastern Brazil, but tracing the path of the huge storms, they determined that the only possible path was across the ocean from Africa. Checking satellite photos, they were even able to distinguish a steady plume of dust and trace its course across the Atlantic. It was well known that Sahara dust had been able to reach Guyana and the Caribbean, and way back in 1846 Darwin had reported a snow of white dust settling on the *Beagle* as it rode the swells off South America, but the idea of the Sahara feeding the Amazon was a new one.

Now the team is studying soil cores to determine what possible effect this process might have had on the genesis of the forests, and they are trying to assess its effect on marine organisms, particularly planktons, that thrive in the paths of the storms.

Once, Lake Chad was ten times its current size, and the sediments around it are rich in diatomaceous earths and salts. From its present desolation comes the dense life of the rainforest.

THE SOIL MAN

HANS JENNY, who died in 1992 at more than ninety years of age, was one of the great scientists of this era. Were he not a soil man, Jenny's achievement might have ranked him with the pioneer ecologist Henry Cowles or with the physicist Clerk Maxwell, but his work has been comparatively buried in the obscurity of his profession. The head of the University of Chicago's venerable department of ecology, asked to comment on Jenny's contribution, remarked, "Never heard of him." And not long ago, at a University of California cocktail party, Jenny told a new acquaintance he was a soil scientist only to be asked about garden tomato plants. This is like asking Toscanini to tune your guitar.

Such incidents are doubly unjust. Jenny's formula for soil genesis was one of the pioneering works of ecosystems ecology. It is an archetype of whole-systems thinking. It was Hans Jenny who gave the picture a new dimension, with his 1941 formulation of soil process that distinctly linked "deep time" and our time. For Cowles, twenty thousand years was old. Jenny looked as deep into time as the age of the oldest soil he could find: more than half a million years. By including the factors time and parent material in his equation, along with climate, slope, and organisms, he was able to link biology, geology, and duration in a single mathematical relationship. The equation, called CLORPT, describes a set of feedback loops that maintain a living system.

Ecologists sometimes receive unusual monuments. Hans Jenny kept a piece of what may well become his own monument in the lower-right-hand drawer of his desk, which is often where a man keeps what he most values and/or fears. The ninety-two-year-old Jenny pulled open that sacred drawer in his University of California office in Berkeley. "I want to show you something," he said in his clipped Swiss accent.

He handed me a white lump the size of a softball, but chalky, as though he had been saving blackboard dust for half a century. "Hold it," he said. "See how heavy it is." It sank in my hand like a shotput. "Now that has hardly any organic matter in it, hardly any nutrients," he observed slowly, choosing his words. "But it is certainly a soil."

It was also my introduction to the pygmy forest, and Jenny's understanding of ecosystem evolution.

The heavy chalky lump I hefted was, he told me, ancient, Methuselan. With a pH just this side of lemon juice and a subsoil as hard as a frying pan, the soil it came from grew nothing but a few stunted pines and heath plants—acid lovers like manzanita—many of which sickened and died, still standing.

What calamity brought this about, I asked him.

Time, nothing but time. More than half a million years of sitting out in the rain.

Jenny sat regarding the bagged lump with the eyes of a proud father. It represented, I believe, his excursion to the edge of human time, his adventure in truth. For more than two decades since his retirement from active teaching, he had pursued research on this soil, trying to establish how it had come to be.

What most startled me was that the soil was not from a parched desert but from the area widely regarded as unmatched for scenic beauty on the whole California coast. This white dust was from Mendocino.

Several months later, Jenny and I were bound north from Berkeley in his station wagon to see this soil in its place. In the back of the car rattled cans of tunafish and fruit and a jar of coffee, together with a tackle box full of little augers, a hammer, chicken wire, white envelopes, and labels. Somewhere in the wine country of Sonoma, we scooted off onto a side road, and he told me to stop. "What do you see?" he asked.

I saw a hillside with some live oaks, some vines, and some pine trees.

"Well, why," he continued imperiously, "does the oak grow over here and then the pine grow there?"

Before I could answer, he was out of the car and swinging off along the roadside. He stooped to pick up some small stones and returned to me. In one palm he held a blondish sandstone the color of dried grass; in the other, a handful of small, sharp-edged, friable chunks of a deep-green rock called serpentine.

Here in my own native ground, he had shown me a key to the landscape that I had never noticed. Other factors being equal, a soil derived from sandstone will support oaks and vines; a soil that comes from serpentine is covered with scraggly digger pine.

As we drove west over the coastal hill toward the sun that hovered above a bed of fog, Jenny explained to me that the pygmy forest soil that we were going to see had deeply affected his idea of ecosystem evolution and caused him to question the unspoken idea of much environmental thinking: that, properly treated, nature balances herself in a way that is beneficial to man.

That evening, in the musty-smelling clapboard farmhouse that has been his base for decades, he described the staircase of five marine terraces that step up the hill from the town of Mendocino to an altitude of more than six hundred feet above sea level. The top three terraces contain areas of pygmy forest with its dust-and-iron soil. The underlying rock on which the whole landscape is based is a graywacke sandstone, laid down fast in a deep sea trench about 150 million years ago. It is what geologists call a "poorly sorted" stuff, which means it contains the whole range of mineral elements that, once weathered and made soluble in soil, contribute to plant growth.

Jenny's picture of soil evolution reaches to a time in the middle Pleistocene, between a half million and a million years ago, when the sea level rose, responding to melting glaciers farther north. As the water rose, waves cut a shelf in the graywacke. Then the glaciers returned, and the sea level fell, the receding water leaving a layer of stones, gravel, and sand on the now-exposed shelf. The glaciers retreated again, and sea level rose, but at the same time tectonic forces of compression along the San Andreas Fault, at the junction of the North American and Pacific plates, lifted this first shelf above the reach of the waves. So the sea began to cut a new shelf, at a lower level than the first. Over the course of the Pleistocene and into the Holocene, this to-ing and fro-ing continues, until like a gigantic escalator the landscape had unfolded at least five shelves, each made of roughly the same parent materials and each slightly higher and farther inland than the next. The oldest is perhaps three quarters of a million years old; the youngest is still waiting to be born.

Overlaid on this stately sequence of emerging shelves is a corresponding set of fore-edge dunes. Each was the result of sand blowing up from the beach and becoming piled on the front edge of the most recently elevated terrace. Thus, each shelf has two ages, one belonging to its own materials and the other belonging to the dune that was later blown up onto it.

This, Jenny asserted, was the history of the landscape that made it possible to study the effect of time on the pygmy forest soil. Though fascinated, I found it difficult to imagine why this sequence of events should result in a soil barely able to support life, as Jenny contended it had. The claim became doubly hard to swallow the next morning, when he drove me to the Mendocino headlands.

Down below us on the beach, we could see sloshing beneath the incoming waves the pebbles, gravel, and chunks of tough graywacke that would be the raw material for a future soil on a still unformed terrace. We were standing at the top of the headland, right at the edge. It looked barren enough on the steep slope of the headland, where lupine, iceplant, and pinks were holding on in the crevices. But when Jenny turned around, he plunged his hands into an exposed soil profile that was as black and as rich-looking as the Iowa prairies. The soil of the lowest terrace, on which we were standing, is a grassland soil of the sort called "mollisol," related to the soils of the Midwestern prairies as well as to the famous Russian chernozem. The mutual action of organic acids and the chemical and mechanical weathering of potassium- and calcium-rich graywacke sands had turned the sand to clays, and along the edges of the clay particles clung the blackish organic complexes of humus substances. The soil supported a magnificent meadow flora of nodding onion, wild iris, lupine, buttercup, bunch grasses, strawberry, yarrow, and many more species. How was I to believe that this soil was a younger brother to one not three miles distant that supported more lichens than plants?

We drove up to the next terrace. Not only was there no more prairie, we could not even see the prairie because we were deep in a mature forest dominated by majestic redwoods and Douglas fir, with thick stands of rhododendrons, salals, and other ericaceous (of the heather family) plants. The dense growth made for tough walking, but the acid odor of the conifers, together with the rhododendron flowers and the waxy berries of the salal, were more than adequate compensation for the trouble.

Fifty thousand years ago, if Jenny was right, the place where we were standing had itself been a prairie by the edge of the sea. Propelled by tectonic forces, it was now a little higher and much older than its prairie brother below. The soil here was more weathered, and the soil horizons had become thicker and more distinct. At the bottom of this soil could still be seen the same sands and gravels of the original beach material, but now they were so worn that you could crumble them in your hands.

At the next step in the staircase, Jenny showed me a forest dominated by a towering bishop pine. At first glance, the older soil here seemed to support a plant community as robust as that on the shelf below. But there was less diversity of species, and the forest was less dense. The trees were sending their roots deep, in search of the nutrients that were being leached all the way to the water table, in which they might finally be lost. In this soil, the top gray horizon had thickened, as weathering leached more and more minerals from the surface soil, leaving only the grayish, resistant quartz and kaolinite. The iron drops of the lower horizon had grown more numerous, and some of them were cemented into clusters.

From the aboveground life on the three shelves, I would never have guessed that these landscapes were closely related. Yet in the soil horizons, I could clearly grasp their common origin and their evolution. We drove up over the hump onto the next terrace and took a dirt road onto property that belongs to the University of California. Jenny and his wife struggled for years to get anyone interested in preserving this landscape, and eventually won a state park for one strip of the staircase we had ascended as well as this U.C. study plot in the pygmy forest. "People don't regard soils as beautiful," he lamented, "so it's hard to argue why they should be preserved." He is only half kidding, maybe even serious, when he suggests turning the selenium-tainted wetlands of Central California into Selenium State Park. The tainting, after all, is the result of the natural leaching of a trace element when irrigation water is poured over the soils.

Leaving the car beside a rut, we walked through a scrubby path of conifers into another world. The tallest trees were scarcely taller than a man, and many rose no higher than our waists, though they were decades or even centuries old. Their trunks were as slender as a mummy's wrists. Here grew dwarfed, twisted versions of the plants and cypresses on lower steps of the staircase. A few were endemic species, found nowhere else in the world. Almost a quarter of the area was bare ground or covered with yellow-green lichens. Thirty percent of the trees were dead or dying. When they perished, they remained standing, rotting in place.

From above, who could ever say what had caused this apparent catastrophe? Standing with Jenny in an eight-foot-deep soil trench, I could see the answer. The very bottom of this soil, where it met the unaltered graywacke sliced flat by the rising waves of a million years ago, was the same beach sand as on the other terraces, but the horizons above were the sclerotic developments of the processes that were still in full swing on the levels closer to the beach. Beneath a slender, gray-brown top layer was a bone-white horizon at least one foot thick. There were no free nutrients left in this layer, only the most resistant, insoluble quartzes. It was from this layer that Jenny's lump of dust had come. The metallic elements leached by millennia of rain from that graying surface now formed not teardrops or even clusters of red-brown knots, but a solid, unbreakable hardpan horizon, in places more than three feet thick. To get a piece of it, you had to hit hard with a hammer more than once.

Little could live atop this white-and-red soil. Jenny and his students were still doing experiments to grow other plants here—poppies, grasses, anything. On unaltered soil, nothing at all would emerge. If the soil had been amended with a nitrogen fertilizer, the plants would sprout, use up the fertilizer, and keel over dead.

There was not, in fact, a lack of nitrogen. Because the hardpan prevented drainage, during the winter rains the whole forest was awash in a coffee-brown liquid of water mixed with humus substances rich in nitrogen. But the nutrients derived from the mineral world, particularly phosphorus, were virtually lacking, having leached away, or having been locked up in compounds that were very hard to break and therefore useless to plants. Furthermore, because the soil surface was cut off from the depth, the acidity of the soil had built to such a level that few soil microbes could survive, so that plants growing here would have little access to the nutrients usually produced or converted into usable form by such organisms.

As Jenny and I stood on the slope leading into the soil trench, the old man was filled with the delight of his knowledge. "You see," he said stooping, "the soil down at the bottom of the trench, *beneath* the hardpan, has more nutrients than the topsoil. If we find a living seedling, it will likely be down there." Down we went, though it occurred to me there was some chance that his body, so old he could scarcely keep his head erect on his neck, would never leave the trench.

We scrabbled about in the red subsoil until we found a tiny seedling, perhaps two centimeters tall, of the endemic bolander pine.

"Now that is a seedling two years old," he said exultantly. Anywhere else, it should have been inches, not centimeters, high.

When he stood up again in the trench, his eyes, like mine, rose just a few inches above the soil surface. "It's beautiful, isn't it," he breathed, looking out over the miniature, contorted landscape.

I wondered what he could mean by that. He had shown me that an apparently random assemblage of landscape features had a deep underlying order whereby prairie turned to forest, turned to pygmy forest. Soils, under the influence of time, were largely responsible for the changes.

"Your ideas are beautiful," I told him, "but this landscape is frightening."

Jenny had stood me on the boundary between deep time and our time. I could not avoid the feeling that we are just one experiment in a more ancient world. "What does nature have in mind in making soils?" he had once asked. In light of the pygmy forest, he could not answer with the communitarian optimism of a sanguine environmentalist. Over the long haul, nature was not in the business of making pleasant places for people to live. Quite the contrary, it seemed interested in pushing the limits of the relationship between the organic and inorganic realms, producing new experiments like the scraggly bolander pine.

Yet to Jenny this is a cause for wonder, not despair. His study provokes the most difficult beauty, the one that we would often as soon leave buried in that lower-right-hand drawer. It is a beauty that admits the underground, the underworld, the soil, the dirt, heat, decay, cold, smells, soluble metals. It may be a hideous, misshapen, twisted, threatening thing on the surface, but musically beautiful in the laws by which it lives. According to the poet Rilke, Orpheus never sang so sweetly as when he went to the land of the dead.

Certainly, the scientist could accept such a characterization of what he meant by "beautiful." When I said I thought the landscape frightening, Jenny did not even turn his head. He kept looking out across the dwarf forest floor, his nose practically resting on the edge of the trench, and growled, smiling, "Ah! You must look with fox's eyes."

INTO THE PITS

I ROUNDED THE BEND OF THE FIELD just as the autumn sun came over the eastern hills, racing over the sorghum and lighting the rough edge of the pine forest. There before me I could see eight young men and women in muddy clothing. They were seated, kneeling, standing, or pacing beside an open pit.

A black woman in an Auburn University T-shirt was spraying water from a squeeze bottle onto a reddish lump of soil. A square-jawed man with a blond ponytail looked intently at a smudge of dirt that he was working with a forefinger into the palm of his hand. A broad-shouldered, florid man in a John Deere cap broke off a piece of soil and inserted it between his cheek and gums. The remaining part he held to his nose while he inhaled. I felt that I had come upon a latter-day school of alchemists, seeking with all of their senses to decipher the material world.

As I watched, someone said in a loud voice. "Ready? Change!"

Another eight young people came boiling out of the deep trench where they had been concealed, each of them bearing trays or cups full of soil. The original eight drew their knives and rushed down the sloping entry ramp into the ground.

I walked closer to see what they were going to do. The pits were at least twelve feet long, five feet across, and six feet deep, more like the entry to an underground barrow than a simple fosse, or ditch. Nevertheless, because of their nervous faces and the drawn weapons, I had the impression that I was going to witness a gangland remake of the fight between Hamlet and Laertes in Ophelia's grave.

Instead of attacking each other, however, the eight faced the walls of the pit. They scored it, touched it, rubbed it, wet it, smoothed it, and cut off hunks and chunks and bits of it. They stared earnestly at the places where one color faded into another, or where mottles appeared, or where suddenly the soil gave way to a rotten whitish rock that crumbled between the fingers.

Agronomists, 4-H kids, and F.F.A. members all hold contests to measure and identify soils, but few are as fervid as the teams of college students in the southeast. Twelve teams from five states appeared for the southeast regionals of the National Soil Judging Competition in the North Carolina Piedmont. Just at dawn on the day of the contest, they'd been driven to a site where four pits that they had never seen before awaited them. Each of the more than seventy-five students would have a little better than half an hour per pit to examine, measure, and accurately describe the four soils.

It was comical to see them descend like pirates into the pits, then reemerge and seat themselves on the damp, matted grass, where straightaway they were transformed into contemplatives, interrogating their soils with every sense, not excluding taste. I have seldom seen a funnier or more graceful sporting event.

After they were done, I went alone into each pit. It was like entering the theater of a twelve-thousand-year-long play. The four holes were no more than five hundred yards apart, dug into the edge of the piedmont where the remains of ancient seas are jumbled together with older metamorphic bedrock.

Nothing in these soils was not alive. Nothing in the profiles was not moving. The tendrils of the roots made blond channels in the whitish yellow layers, from which for millennia water had leached the iron. Deeper in the profile, the iron had gathered into clumps of subterranean rust, freckling the soil against a base of ochre clay that was greasy to the touch. In one hole, the lower horizons had divided into deep red and golden layers, like cirrus clouds on an autumn afternoon. In another, a solid boulder of granite flaked and crumbled at a touch into sugary granules. In another, all that was left of the bedrock were shiny flakes of mica that sparkled in bloodred clays. In a fourth, a layer of pebbles had been lifted up atop the red clay, like an offering of the earth to the sun.

I thought of Hans Jenny, who now has been dead for more than a year, and of what he had said to me about looking with fox's eyes. Since that time at the pygmy forest, I'd often wondered what he meant. In these pits, following these knife-wielding students, I learned the answer: Everywhere he looks, the fox finds life.

THE EARTH FOR JEFFERSON AND ADAMS

The United States is a democracy; it does not accomplish its ends by handing down decrees from above, but by the initiative and the consent of the citizens, who must first know what they want and how to achieve it.

—*Soils and Men*, THE 1938 YEARBOOK OF THE DEPARTMENT OF AGRICULTURE

WHEN I WAS A BOY, it seemed like new things were being discovered every day. On the air force base where we lived, for example, there were wedge-shaped planes that made sounds never before heard on Earth. For years, I proudly kept a slender roll of paper given me by a pilot, on which a stylus had traced the sound of his sonic boom. In the history of sonic booms, the one that I possessed was certainly among the first one hundred. How many tens of thousands of those noises have disturbed the sky in the forty years since? And to what little purpose?

The sonic boom gives a good picture of the wave of technological innovation that swept the world right after World War Two. It was sudden, fast, very loud, and in large measure, served to dirty the atmosphere and to stir fear in the mind. And it was really exciting.

But the heat of the affair is over. I can't remember why I was so moved by it. Of course, the scientists are still there, spending ever larger amounts of money to bang out ever smaller and more ephemeral particles, while ignoring the simple anomalies of quantum mechanics, which make irrelevant gibberish of their costly speculations. Some journalist has the temerity to compare the vastly expensive and useless supercollider to the modest organ on which Bach composed, when a more appropriate comparison might be to the impregnable fortresses, which were to become mass graves, built by the Byzantine Emperor Justinian in a disastrous effort to extend his frontier with Persia.

Even those who have a vision tend to look among the usual suspects. If one branch of science is moribund, then look to the progress in computers and video. Imagine an interactive future—quantum mechanics writ large—in which perceiver and perceived continually co-create each other.

But above all, do not look behind you. A wall seems to separate us from the aspirations of our forefathers. It is a wall made of garbage, in which the history of the world and of our republic mixes with reruns, suburban streets, the fishlike bodies of detergent bottles, the compacted wads of a million diapers, the mangled tapestry of disused tricycles. All around the cities and their suburbs are rings of trash that cut us off from the garden as effectively as any angel with a flaming sword.

We live in the garbage heaps of culture. But we have drawn the wrong conclusions from this fact. Dump heaps have not properly been regarded as very savory things by us. We have hidden them in the backwoods and sought to cover them over with enough earth that they resemble little hillocks. Yet, rightly treated, the dump heap becomes the compost pile. Ideas buried there devour each other, and get hot with the heat of metabolism. Out of the transformed heap spring healthy new plants from seeds that we had long since forgotten.

I propose that we compost Jefferson and Adams, and plant our own meditations in their fertile earth.

John Adams and Thomas Jefferson, the second and third Presidents of the United States, died within an hour of each other. Adams's last words were "Jefferson lives." The death of the pair was the culminating event of one of the great friendships in modern history and the final stop to five decades of correspondence that constitutes a great work of American literature.

Jefferson and Adams were for each other what neither could be for himself. The tall angular nervous Jefferson and the squat pugnacious Adams are archetypal poles in American leadership, but in their correspondence, the two men composted their differences. Like the soil, they transformed the two detrital realms of their lesser selves into a more fertile whole. Real friendship is a body in nature, invisible to the eyes maybe, but with a power to renew even the commonest idea.

Jefferson is known as the Agrarian, the man who recommended a dispersed population and a healthy yeomanry as the foundation of the republic. As many mistaken ideas have been drawn from Jefferson's agrarianism as have misguided notions been derived from Darwin's principles of natural selection. Every plaid-shirted back-to-the-lander cites him as a forebear, but the truth is that Jefferson's farms of several hundred acres were not well managed, because Jefferson himself was so seldom upon them and his overseers did not serve him well.

His enthusiasms for experimenting with new useful plants were many, but largely unsuccessful. He tried olives and Egyptian rice, mulberries and silkworms, sugar maples, grapevines, and oranges. Not a single one flourished. A young granddaughter wrote with amusement of the Seville orange trees they'd planted at Monticello: "Your orange trees come on very well as to their looks," she wrote to her grandfather in Washington, "but I never saw such short little things in my life. They are near eighteen months old and they are not as high (any of them) as my hand is long."

Like the Romantic poet Byron, Jefferson longed for a Mediterranean kind of plenty and was unwilling to admit that his climate might deny it to him. *Jefferson's Garden Book,* compiled from notebooks and correspondence by Morris K. Betts, is a rhythmic and seasonal poem of excitements and frosts: "May 5—A frost which destroyed almost everything, it killed the wheat, rye, corn, many tobacco plants, and even large saplings . . . all the shoots of the vines . . . Oct 12—A frost of four or five weeks' duration, the earth being frozen like a rock the whole time. This killed all the olives. . . . April 17—The forest totally blasted, every leaf being killed on the hardiest trees. . . . Nov. 28—It is so cold that the freezing of the ink on the point of my pen renders it difficult to write. . . ." For all that the New World seemed lush, it was resistant to Jefferson's agricultural dreaming.

But his willingness to experiment was his strength, too. Jefferson was little concerned whether or not he was regarded as a fool. He was at his best and his worst at Monticello. Jefferson was among the first in the United States to recommend full crop rotations, with periods of planting in legumes and periods of grazing, to renew the soil. He created an elaborate scheme to accomplish this, field by field. He opposed as destructive the practice of bare fallowing, which left the land open to erosion, and he was an early advocate of contour plowing, working across the slope so that the furrows would hold water, not channel it into runnels. His recommendations were virtually the same as those adopted by the Soil Conservation Service 150 years later.

But Jefferson's agrarian vision was not robust enough. His system of rotation never was put into practice, because the economy defeated it. Always cash-poor, Jefferson depended on his small manufactures and service businesses, like his naillery and his mill, to provide the wherewithal for his large farm. When these failed, as they all did eventually, he was hard put to make up the loss.

He turned to wheat and to tobacco, the single most soil-exhausting crop known to the farming of his day. To a friend he confessed, "The high price of tobacco, which is likely to continue for some short time, has tempted me to go entirely into that culture, and in the meantime, my farming schemes are in abeyance." He was unequal to the interface of money with his farm, so he was unable to prevent, even on his own ground, what he complained of in Virginia at large: "We would sooner buy a new acre of land, than manure an old one."

In this respect, he was the opposite of his friend and political rival, John Adams. For all the history books tell, one would expect that Adams, the Yankee and a leader of the Federalists, would be a friend of urban manufacturers and an enemy of the agrarian vision. But Adams was in fact a fine and successful farmer, at a scale appropriate to his means.

What is more, he understood the importance of a robust farming community to the health of the nation's cities. It was he, not Jefferson, who pushed through the Continental Congress of 1774 a resolution to establish in all the colonies Societies for the Promotion of Agriculture.

He was so proud of this achievement that he noted the event in his journal, specifically claiming authorship of the idea.

His motive was both to make the colonies self-sufficient with respect to food and clothing and to increase the wealth of the incipient nation. He envisioned a healthy and local relationship between farms and factories. He was the first of the friends to suggest the wisdom of importing Merino sheep and improving the breed for domestic woolens. If one looks for the father of the county fair in America—and of all local initiatives to advance farming on an appropriate scale—one must look as much to Adams as to Jefferson.

Adams also understood the soil, viscerally and emotionally, as his friend did not. The son of a farmer and inheritor of forty acres of Massachusetts soil, the New Englander was ever proud of his manure heaps. To Charles Francis Adams, the cultured scion who as his ancestor's editor also excised the references to lovely women, this was *de trop*. So we have inherited a lawyerly idea of the second President and have entirely missed the hearty life of him.

Jefferson, the "agrarian," was an inventor and a dilettante. Adams, the "federalist," was a farmer and a compost man. Go figure. Each man contained the other's shadow. Jefferson had a heart that rose into his mind; Adams's heart sank deep into his instinctive life.

As I write, I am looking out the window at two old peach trees whose fruit has turned overnight from yellow-green to orange-red. This is only a rented country house, and I am only a harried writer. The farmers are long gone from this place, but the soil is faithful. Five or six peaches are already lying on the ground, sinking into the tufted grass.

What if I bothered to learn from the past? What if I regarded it as significant that this pair of friends died on the same day, July 4, 1826, on the fiftieth anniversary of the republic they founded? What if I composted the differences between Jefferson and Adams, and sought in their union and transformation some clue to a right relationship to the land and to the stuff of my own life?

STARDUST

EMERSON WROTE THAT IF THERE WERE GOOD MEN, we would not go into such raptures over nature. He cited an old proverb, one I've never heard elsewhere: "When the king is in the palace, no one looks at the walls."

By "king" he did not mean someone unknown to us. He meant himself and each one of us. After five years of work on the soil—looking at these "walls" for their beauty, usefulness, strength—I have come to the conclusion that I ought to start all over again. I ought to write about the man the soil suggests.

Hans Jenny is as close as I have found. He was a man of deep integrity. With seven decades of hard-won knowledge, he confessed his ignorance. He insisted on seeing whole, when others made a virtue of seeing in slices. He knew science as a form of prayer.

Even Hans is not enough, though. Each of us is made of stardust, as my boss, Jim Morton, preaches every year. We have each, then, the stuff in us and the bound-up energy that might launch a beam of light.

Soil is only the darkest and coldest of all living things. The most widespread. And the most receptive. Warmed, it blooms. So may I in my darkest moments be attentive to the penetrating rays of the sun that finds the seed.

Work, motion, life. All rise from the dirt and stand upon it as on a launching pad. At the outer edge of the atmosphere, the thin air continually gives off hydrogen ions that join the solar wind. To what end and to what stars might this lightest, quickest dust be bound?